Splunk 9.x Enterprise Certified Admin Guide

Ace the Splunk Enterprise Certified Admin exam with the help of this comprehensive prep guide

Srikanth Yarlagadda

BIRMINGHAM—MUMBAI

Splunk 9.x Enterprise Certified Admin Guide

Group Product Manager: Reshma Raman

Publishing Product Manager: Apeksha Shetty

Content Development Editor: Shreya Moharir

Technical Editor: Kavyashree K. S.

Copy Editor: Safis Editing

Language Support Editor: Safis Editing

Project Coordinator: Farheen Fathima

Proofreader: Safis Editing

Indexer: Rekha Nair

Production Designer: Shankar Kalbhor

Marketing Coordinator: Nivedita Singh

First published: August 2023

Production reference: 2200923

Published by Packt Publishing Ltd.
Grosvenor House
11 St. Paul's Square
Birmingham, West Midlands B3 1RB
United Kingdom

ISBN 978-1-80323-023-8

www.packtpub.com

To my father, Nageswara Rao Yarlagadda, and my mother, Suseela, whose love and guidance have shaped my world. To my wife, Sushmitha, whose unwavering support fuels my aspirations. To my daughter, Sanvitha, and son, Sathvik, whose laughter fills my days with happiness. You are my strength, my inspiration, and where my heart truly belongs.

- Srikanth Yarlagadda

Contributors

About the author

Srikanth Yarlagadda is a seasoned Splunk Security Consultant, specializing in maximizing the potential of the Splunk platform to fortify an organization's cybersecurity stance. His primary responsibilities encompass harnessing the capabilities of Splunk Enterprise and its **Security Information and Event Management (SIEM)** and **Security Orchestration, Automation, and Response (SOAR)** functionalities. Srikanth excels in monitoring, detecting, tuning, and automating security threat scenarios for **Security Operations Center (SOC)** teams.

With a robust background, Srikanth boasts extensive experience in administering Splunk Enterprise. He has adeptly managed system administration tasks and expertly orchestrated data onboarding from diverse sources. His proficiency extends to developing a wide array of business use cases within the telecom domain, encapsulating the power of Splunk. Srikanth has obtained the following official Splunk certifications: Splunk Enterprise Architect, Splunk Admin, Enterprise Security Admin, and Splunk SOAR Developer.

In his earlier ventures, Srikanth's IT journey took him through various technology domains. He took on roles as a skilled Java developer and an adept SOA developer and even ventured into the realm of telecoms protocol implementations using **Oracle Communications Services Gatekeeper (OCSG)** and Oracle SOA Suite.

Srikanth's expertise is further evidenced by his active engagement as a SplunkTrust member, a testament to his contributions and dedication. His participation is frequent across Splunk's vibrant community forums, where he shares insights and collaborates with fellow enthusiasts.

I am immensely grateful to the organizations that have extended their trust and provided me with opportunities. Their belief in my abilities has propelled me to the stage I stand on today.

About the reviewer

Trayton White, self-proclaimed "World's Okayest Splunk Wizard," has spent his career tinkering with systems in a variety of industries. Most recently, he worked on new Splunk offerings by testing them and developing novel use cases the technology helps solve. By adopting a "let's try it and see" approach, he has developed in-depth knowledge of the inner workings of Splunk and enjoys sharing this information with others.

Table of Contents

3

Users, Roles, and Authentication in Splunk   45

4

Splunk Forwarder Management   59

5

Splunk Index Management   75

6

Splunk Configuration Files   89

7

Exploring Distributed Search   105

Part 2: Splunk Data Administration

8

9

10

11

Field Extractions and Lookups 183

12

Self-Assessment Mock Exam 211

Index 227

Other Books You May Enjoy 234

Preface

Welcome to your preparation guide for acing the Splunk Enterprise Certified Admin exam—an essential step in mastering the world of data management. In this concise yet comprehensive handbook, we will equip you with the knowledge and strategies needed to confidently navigate the certification journey and emerge as a certified Splunk Enterprise Admin.

In today's data-driven world, harnessing information is key. Welcome to *Splunk 9.x Enterprise Certified Admin Guide*. This book is your path to unlocking Splunk® Enterprise's full power—a top platform that helps businesses turn raw data into valuable insights.

Who this book is for

The audience for this book includes data professionals interested in becoming certified Splunk administrators. Additionally, the book is suitable for data analysts, IT professionals, system administrators, Splunk users, and security analysts who work with data and are interested in leveraging the power of Splunk to make sense of it.

What this book covers

Chapter 1, Getting Started with the Splunk Enterprise Certified Admin Exam, serves as an introduction to the Splunk Enterprise Certified Admin Exam and provides an overview of the key concepts and skills that the exam covers. It prepares you for the subsequent chapters by setting the context for the various administrative tasks discussed throughout the book.

Chapter 2, Splunk License Management, explains Splunk licensing, including different license types and how to manage and monitor license usage. It covers the importance of proper license management to ensure optimal usage of Splunk's features and capabilities.

Chapter 3, Users, Roles, and Authentication in Splunk, focuses on user management, roles, and authentication mechanisms within Splunk. It covers creating and managing user accounts, assigning appropriate roles and permissions, and configuring authentication methods to ensure secure access to the Splunk environment.

Chapter 4, Splunk Forwarder Management, delves into the management of Splunk forwarders, which are used to collect and forward data to the Splunk indexer. It discusses the installation, configuration, and management of forwarders using the deployment server.

Chapter 5, *Splunk Index Management*, introduces the concept of indexes in Splunk, which are used to store and manage data. This chapter covers creating and managing indexes, configuring data retention policies, and optimizing index settings for efficient data storage and retrieval.

Chapter 6, *Splunk Configuration Files*, provides valuable insights into Splunk's configuration files, which play a pivotal role in customizing and fine-tuning the Splunk environment. This chapter delves into various configuration files, explores search-time and index-time precedence, and provides guidance on troubleshooting using the `btool` command.

Chapter 7, *Exploring Distributed Search*, is the final chapter of *Part 1*. It delves into Splunk's distributed search abilities, which entails searching and analyzing data across various Splunk instances, including an introduction to clustering. This chapter addresses configuring distributed search, examining the knowledge bundle, and making adjustments to minimize its size.

Chapter 8, *Getting Data In*, serves as an introduction to ingesting data into Splunk. It explores various methods and sources for data input, helping you understand how to collect and prepare data for effective analysis.

Chapter 9, *Configuring Splunk Data Inputs*, guides you through the process of setting up data inputs in Splunk. You'll learn how to configure methods such as monitoring files and directories, network inputs, scripted inputs, **HTTP Event Collector** (**HEC**), and Windows inputs. These steps ensure a seamless data flow from various sources into your Splunk instance.

Chapter 10, *Data Parsing and Transformation*, shifts the focus to data manipulation. You'll discover techniques for parsing raw data and transforming it into a structured format, enabling meaningful analysis and insights.

Chapter 11, *Field Extractions and Lookups*, explores advanced data processing, focusing on search-time and index-time field extractions to uncover valuable information from raw data. It also delves into the use of lookups to enrich your data with additional context.

Chapter 12, *Self-Assessment Mock Exam*, reinforces your learning with a self-assessment mock exam. It provides practice questions and scenarios to gauge your comprehension of the concepts covered in *Part 1* and *Part 2* of the book.

To get the most out of this book

To understand the concepts in this book, you will need fundamental Splunk user skills, including basic search proficiency and an understanding of knowledge objects and fields. Additionally, familiarity with Windows and Linux operating systems is essential. If you aim to take the Splunk Enterprise Admin certification exam, acquiring the Splunk Core Power User certification is a prerequisite.

Software/hardware covered in the book	Operating system requirements
Splunk Enterprise 9.x	Windows 10/11 or Linux

If you are using the digital version of this book, we advise you to type the code yourself or access the code from the book's GitHub repository (a link is available in the next section). Doing so will help you avoid any potential errors related to the copying and pasting of code.

Download the example code files

You can download the example code files for this book from GitHub at `https://github.com/PacktPublishing/Splunk-9.x-Enterprise-Certified-Admin-Guide`. If there's an update to the code, it will be updated in the GitHub repository.

We also have other code bundles from our rich catalog of books and videos available at `https://github.com/PacktPublishing/`. Check them out!

Conventions used

There are a number of text conventions used throughout this book.

`Code in text`: Indicates configuration file names, file extensions, pathnames ,Splunk installation and important directory locations, Splunk CLI commands and options, stanza names and configuration settings, . Here is an example: "The location of Splunk system-wide configuration on Unix systems is `$SPLUNK_HOME/etc/system/[default|local]`."

A block of code is set as follows:

```
## inputs.conf *nix-style File monitor stanza
[monitor:///var/log/application/*.log]
sourcetype = application_logs
index = dev_app
disabled = false
```

Any command-line input or output is written as follows:

```
./splunk btool check
```

Bold: Indicates a new term, an important word, or words that you see onscreen. For instance, words in menus or dialog boxes appear in **bold**. Here is an example: "Select **Settings** and then click on the **Distributed Search** menu item."

> **Tips or important notes**
> Appear like this.

Get in touch

Feedback from our readers is always welcome.

General feedback: If you have questions about any aspect of this book, email us at customercare@ packtpub.com and mention the book title in the subject of your message.

Errata: Although we have taken every care to ensure the accuracy of our content, mistakes do happen. If you have found a mistake in this book, we would be grateful if you would report this to us. Please visit www.packtpub.com/support/errata and fill in the form.

Piracy: If you come across any illegal copies of our works in any form on the internet, we would be grateful if you would provide us with the location address or website name. Please contact us at copyright@packt.com with a link to the material.

If you are interested in becoming an author: If there is a topic that you have expertise in and you are interested in either writing or contributing to a book, please visit authors.packtpub.com.

Share Your Thoughts

Once you've read *Splunk 9.x Enterprise Certified Admin Guide*, we'd love to hear your thoughts! Scan the QR code below to go straight to the Amazon review page for this book and share your feedback.

https://packt.link/r/1-803-23023-1

Your review is important to us and the tech community and will help us make sure we're delivering excellent quality content.

Download a free PDF copy of this book

Thanks for purchasing this book!

Do you like to read on the go but are unable to carry your print books everywhere?

Is your eBook purchase not compatible with the device of your choice?

Don't worry, now with every Packt book you get a DRM-free PDF version of that book at no cost.

Read anywhere, any place, on any device. Search, copy, and paste code from your favorite technical books directly into your application.

The perks don't stop there, you can get exclusive access to discounts, newsletters, and great free content in your inbox daily

Follow these simple steps to get the benefits:

1.   Scan the QR code or visit the link below

https://packt.link/free-ebook/9781803230238

2.   Submit your proof of purchase

3.   That's it! We'll send your free PDF and other benefits to your email directly

Part 1:
Splunk System Administration

The initial part of this *Splunk 9.x Enterprise Certified Admin Guide* provides a comprehensive introduction to the core aspects of running Splunk. These chapters provide a robust starting point for effectively managing and configuring Splunk Enterprise. *Part 1* establishes a strong foundation for aspiring adept Splunk administrators, covering key elements such as the exam prerequisites, important Splunk features, proven setup methods, license management, user access control, forwarder management, index management, exploring configuration files, and implementing distributed search, including clustering fundamentals. These chapters will equip you with essential knowledge and skills to efficiently oversee Splunk Enterprise.

This part contains the following chapters:

- *Chapter 1, Getting Started with the Splunk Enterprise Certified Admin Exam*
- *Chapter 2, Splunk License Management*
- *Chapter 3, Users, Roles, and Authentication in Splunk*
- *Chapter 4, Splunk Forwarder Management*
- *Chapter 5, Splunk Index Management*
- *Chapter 6, Splunk Configuration Files*
- *Chapter 7, Exploring Distributed Search*

1

Getting Started with the Splunk Enterprise Certified Admin Exam

Let's get started with Splunk Enterprise. By the end of this chapter, you should understand what Splunk Enterprise is and its rich set of features and be able to list the Splunk components that work together to get business insights out of data. You will also learn about the installation of standalone Splunk Enterprise in a Windows environment, along with advanced **Splunk Validated Architectures** (**SVAs**) covering all the Splunk components. Throughout the book, you'll often find us using the terms *Splunk Enterprise* and *Splunk* interchangeably. They both refer to the product itself. You will rarely find references to *Splunk Inc.*, which refers to the company that developed and offers the Splunk Enterprise product.

This chapter covers the following topics to get you started:

- Introducing the certification exam
- The weightage of topics in the exam
- Introducing the exam's test pattern
- What is Splunk Enterprise?
- Introducing Splunk 9.x Enterprise features
- Understanding Splunk components
- SVAs
- Splunk installation—standalone
- Self-assessment

Introducing the certification exam

The *Splunk Enterprise Admin* exam is the prerequisite to attain the Splunk Enterprise Certified Admin certification. The exam contains 56 questions that you need to answer in 57 minutes, and you will get an extra 3 minutes to review your answers, bringing the duration of the exam to a total of 60 minutes. Successful candidates will be issued a digital certificate along with Splunk digital badges. In order to be eligible to sit the Splunk Enterprise Admin certification exam, you should have already passed the *Splunk Core Certified Power User* exam and obtained that certification.

The exam tests your knowledge of Splunk Enterprise system administration and Splunk data administration concepts. *Splunk Education* and/or Splunk **Authorized Learning Partners** (ALPs) offer administration courses through instructor-led training along with material, labs, and sample questions. Splunk recommends going through these training sessions. They are paid courses. However, do note that taking part in this training is optional for the admin exam. This book covers both system and data administration concepts along with self-assessment questions on each topic, for you to get ready for the exam.

A Splunk Enterprise system administrator is someone who looks after the Splunk Enterprise platform on a day-to-day basis. This exam tests your knowledge of user management, installation, the configuration of Splunk Enterprise, forwarder management, license management, **search head** (**SH**) management, index creation, indexer management, and monitoring the whole Splunk platform using the **Monitoring Console** (**MC**).

Splunk Enterprise data administrator responsibilities include getting the data into Splunk from various sources, such as data inputs leveraging the **universal forwarder** (**UF**), network inputs, scripted inputs, and **Technology Add-ons** (**TAs**). The data admin ensures the data is correctly broken down into individual events, applying timestamps and setting `sourcetype` and other metadata fields. In addition, they can create knowledge objects required to support other Splunk features for data insights and data retrieval using the Splunk **Search Processing Language** (**SPL**).

The following section explains the weightage of exam questions per topic that are asked.

The weightage of topics in the exam

A list of topics in scope and their weightage has been provided by Splunk in its test blueprint for the admin exam. The topics might be slightly updated by Splunk in the future. At the time of writing this book, these are current and valid for the Splunk Enterprise 9.x Certified Admin exam.

Refer to the latest blueprint prior to booking your exam and find out whether any new concepts have been included. You could try accessing this blueprint using this link: `https://tinyurl.com/36x7apnr`. Otherwise, if the web link changes, look for the blueprint PDF deep link in the *Splunk Certification Exams Study Guide* (`https://www.splunk.com/pdfs/training/splunk-certification-exams-study-guide.pdf`) on the *Splunk Enterprise Certified Admin* page.

Don't be alarmed by the length of the topic list; the topics are covered in thorough detail in the rest of this book, to get you prepared with confidence.

Now that you have an idea of the topics and their weightage, let's understand the exam's test pattern.

Introducing the exam's test pattern

The exam contains 56 questions to be answered in 57 minutes. Each question has at most five options. Some of the questions will have more than one answer, under the *Select all that apply* category. Others are either *true or false* or *single-answer*.

The following are sample questions of the different categories with answers.

True or false category

Q. Splunk Enterprise is only able to store and retrieve text-based data.

    A. True

    B. False

Here, the answer is option A.

Single-answer category

Q. A UF is sending data to `index=linux_os`, which does not exist on the indexer layer. What happens to the data in this scenario?

    A. Since no such index has been configured, the data will be ignored by the indexer

    B. The indexer throws an error message to the UF

    C. A `linux_os` index is automatically created since it did not exist before

    D. The data gets stored in the `lostandfound` index

Here, the answer is option A.

Multiple-choice category

Q. A Splunk admin user has, by default, which capabilities? (Select all that apply)

    A. Admin can install the UF remotely

    B. Admin can create another admin user

    C. Admin can create a custom role for a group of non-admin users

    D. Admin can restart a Splunk SH instance through the GUI

Here, the answers are options B, C, and D.

Let's get started with learning about Splunk Enterprise in the following section.

What is Splunk Enterprise?

Splunk Enterprise is software that collects data from heterogeneous sources and provides interfaces to analyze machine data. Getting to know Splunk Enterprise helps you to choose the right feature for the needs or requirements that will come through while you are working on real-time projects. As an administrator, it is highly expected that you are well aware of these capabilities of Splunk. Key features of this product are explained as follows:

- **Collecting text data**: Splunk Enterprise can only collect and search text data. Non-textual data should not be stored in Splunk Enterprise.

- **Schemaless**: Splunk accepts structured, semi-structured, and unstructured data, and no strict checking of schema compliance is needed.

- **Web, command-line interface (CLI), and REST application programming interface (API) interfaces**: Three standard interfaces are offered by Splunk—web for searching, reporting, alerting, and configuration management; REST API to enable all the web functions through programmatic access; and Splunk CLI for executing system commands, configuring Splunk, and running searches. In general, Splunk Administrators use this interface.

- **Searching, reporting, and alerting**: To query Splunk Enterprise, it has introduced a proprietary SPL, which is used in every interface it offers to retrieve the data from it. Searching enables data retrieval, which could be ad hoc or scheduled to run at a particular time of the day. Reporting involves a reusable search query that is stored and can be scheduled or run on demand. Finally, alerting is a scheduled search and triggers a defined set of actions when a given condition is met—an alert action could involve tasks such as sending an email or executing a script.

- **Anonymizing data**: Data can contain sensitive information, such as **Personally Identifiable Information (PII)** and **Payment Card Industry (PCI)** data. For example, credit card numbers and user phone numbers are highly classified and restricted to only being visible or accessible to a particular group of employees, which is broadly called data sovereignty. To comply with the data standards of an organization, Splunk offers the capability to mask or hide this data during indexing. This prevents users that are querying Splunk Enterprise from discovering this sensitive information. We will study this further in *Chapter 10, Data Parsing and Transformation*, specifically in the *Data Anonymization* section.

- **Scaling from single to distributed deployment**: Splunk Enterprise is designed to accommodate various deployment sizes spanning from individual server configurations to extensive distributed setups. It excels in handling substantial data processing tasks and user support efficiently, even when dealing with data volumes in the realm of petabytes.

- **High availability (HA) and disaster recovery (DR)**: Clustering refers to a group of Splunk instances that work together to enable HA. Multi-site clustering refers to geographically redundant clusters working together for DR. All clustering instances share common configurations through replication.

- **Data collection mechanisms**: Getting data into Splunk is a crucial stage that is a continuous process in large enterprises that comprise various data sources. Splunk provides a UF agent for file monitoring, network inputs, scripted inputs, and **HTTP Event Collector** (**HEC**) for agentless scenarios. Similarly, it provides TAs for collecting data from Linux, Windows, the cloud, CRM, and network devices, and so on. Add-ons are available on the Splunk website (`https://splunkbase.splunk.com`).

- **Monitoring**: The MC application functions as a tool for effectively supervising the Splunk platform. It offers insights into the performance of both standalone and distributed Splunk deployments. The application includes preconfigured alerts and dashboards that can be enabled to ensure proactive monitoring of the platform's overall health and performance.

Let's look at the newly introduced features in version 9.x of Splunk in the following section.

Introducing Splunk Enterprise 9.x features

Splunk Enterprise has evolved over the years and currently stands at version 9.0.3 at the time of writing this book. As it gets more advanced, some of its features become deprecated and new features are added or enhanced. Older versions often reach **end of life** (**EOL**), which means Splunk won't offer support or fix bugs; instead, it advises upgrading to the latest version.

This section covers the important features of Splunk version 8.x that have been carried forward to the latest 9.0 product version, along with new features introduced in the 9.x version. These features are good to be aware of but are not tested in the exam. Feel free to skip this section if you want to:

- **Dashboard Studio**: This provides the necessary tools to create visualizations, such as graphs, charts, and statistical tables, with colors and images. It complements the classic simple XML dashboard that existed in previous versions of Splunk but does not replace it as of version 8.2.6.

- **Federated search**: This is used to search remote Splunk deployments that are outside of the local Splunk deployment. Local SH initiates search requests to remote SH, which acts as a federation provider. Remote deployment could consist of a single SH or cluster.

- **Health report**: Splunk Web has a handy *Health status of Splunk* report that displays the health of Splunk processes in green, red, and yellow states. Selecting each process further drills down into the detailed information. The health report helps admins to get a quick understanding of the platform status, such as I/O wait, ingestion latency, data durability, search lag, disk space, and skipped searches.

- **Durable Search**: Scheduled reports that require the results to be complete for each scheduled run can be enabled to rerun at a later point in time when all the necessary resources are available to finish the job. That's called a durable search. A scheduled report could return partial/incomplete results due to a number of reasons. For example, a search peer might be busy servicing other requests and have exhausted its resources (CPU, memory, and so on). Another scenario is where SH-to-indexer network connectivity is unstable. However, with the **durable search** feature, the scheduler ensures it will rerun the same report at a later point in time for the same window it was supposed to execute and return complete results for. So far, we have gone through the features of the 8.2.x product family. Later sections explain the version 9.0 features.

- **SmartStore Azure Blob support**: SmartStore is a Splunk concept referring to an indexer feature for storing data in remote object storage. In previous versions such as 8.2.X, SmartStore had support for **Amazon Web Services (AWS) Simple Storage Service (S3)** object storage and **Google Cloud Platform's (GCP's) Google Cloud Storage (GCS)**. Starting from 9.0, it also has support for Azure Blob storage.

- **Ingest actions**: Splunk 9.0 introduced Ingest Actions for data administrators with a new UI. It can do data masking, data filtering, and routing through rulesets. It is a cool feature, changing the way data admins traditionally write transform configurations for masking, filtering, and routing. Data could be routed to external S3 object storage and/or to an index. The new **data preview** mode allows uploading sample data of up to 5 GB for live testing.

- **Splunk Assist**: Splunk Assist is an app built for the Splunk cloud offering. It is a fully managed service by Splunk Inc. Starting from version 9.0, the app is available for Splunk Enterprise (on-premises) customers. It provides deep insights to admins regarding Splunk deployment configuration recommendations, evaluating the security posture, making updates to Splunkbase apps, and much more.

- **Cluster Manager (CM) redundancy**: In previous versions such as 8.x.x, there used to be only a single CM for an indexer cluster. Starting with version 9.0, we can configure a second CM and run it in standby mode. Two managers run in an active/standby configuration; when the active manager is down, the standby manager will be active to rescue the whole cluster.

- **Config tracker**: A new internal index, `_configtracker`, has been introduced to track config files and their stanzas, including key-value pairs. This is a cool new feature that helps to troubleshoot config issues and find who, when, and what changed from an audit perspective.

- To go through the complete list of features for previous versions of the 8.x.x family, follow this link and choose the version:

  ```
  https://docs.splunk.com/Documentation/Splunk/8.2.10/ReleaseNotes/
  MeetSplunk
  ```

  Similarly, a full list of 9.0.X features is available here:

  ```
  https://docs.splunk.com/Documentation/Splunk/9.0.3/ReleaseNotes/
  MeetSplunk
  ```

In the next section, we will learn about Splunk Enterprise components.

Understanding Splunk components

Splunk Enterprise has multiple integral components that work together and are primarily divided based on their functions. The list is very comprehensive. A standalone Splunk deployment doesn't require all the components; however, a distributed and highly available deployment requires almost all of them.

A detailed understanding of standalone versus distributed deployment is covered in the following section of this chapter, *Splunk Validated Architectures (SVAs)*. By the end of this section, you will be familiar with two types of Splunk components—namely, processing components and management components.

Processing components

The following are processing components:

- Forwarder
- SH
- Indexer

Let's understand the roles of these components in detail and their association with management components.

Forwarder

As the name suggests, this primarily forwards data from the source to the target indexer. There are two types of forwarders:

- **Universal Forwarder (UF)**
- **Heavy Forwarder (HF)**

UF is a software agent typically installed on the source system where data is being generated. It consists of an input configuration (that is, an `inputs.conf` file) with a list of absolute file paths along with metadata fields such as index and sourcetype. UF is the preferred approach to monitoring and forwarding file contents to designated indexers. By default, UF makes use of the fishbucket process to forward data for indexing exactly once and avoids data duplication through **cyclic redundancy checks (CRCs)** and seek pointers. You will find further information about the additional supported data inputs and detailed explanations about the fishbucket concept in *Chapter 9, Configuring Splunk Data Inputs*.

The following diagram illustrates UF installed on a web server configured to monitor the web server logs and forward them continuously to the indexer as and when the logs get updated:

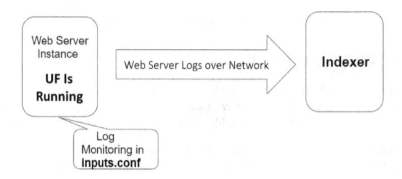

Figure 1.1: UF forwarding web server logs to indexer

Let us now look at SH, which is a critical user-facing processing component in a distributed deployment.

HF is a Splunk Enterprise instance and doesn't require separate binary for installation. It provides an extended feature set compared to a UF. It not only collects and forwards data, but also includes a Splunk user interface for configuration and management. To operate an HF, a forwarder license is required. Typically, an HF is configured in forwarding mode by disabling local data storage. Splunk Add-ons available on Splunkbase can be installed on an HF to facilitate data collection from various sources. This combination of features makes HFs a versatile choice for preprocessing and forwarding data while benefiting from a user-friendly interface.

SH

The SH component is a Splunk Enterprise instance that is dedicated to search management and provides a number of interfaces for users to interact with. The popular interfaces it offers to users are web, CLI, and RESTful API.

Multiple SHs can be grouped together and form a cluster called a **SH cluster** (SHC). Members of an SHC share the same baseline configuration, and jobs are allocated to available members by the SH captain.

In a standalone deployment, a single Splunk Enterprise instance (that is, the same instance) works as both the SH and indexer. In a distributed deployment model, the SH or SHC can submit searches to multiple indexers and consolidate the results returned. The results are stored locally in a dispatch directory located in $SPLUNK_HOME/var/run/splunk/dispatch for later retrieval, and the results will be deleted after the job expires. $SPLUNK_HOME refers to the installation directory where the Splunk software is installed. For example, ad hoc search results (that is, the search job outcome) are retained for 10 minutes in the dispatch directory, which will be removed after the job expires by a process called the dispatch reaper, which runs every 30 seconds.

SH stores search-time knowledge objects that work directly on raw data and/or fields being returned from the indexer—for example, knowledge objects such as field extractions, alerts, reports, dashboards, and macros are categorized as search-time knowledge objects in Splunk.

The following diagram illustrates a distributed deployment configuration featuring a single dedicated SH that communicates with three separate indexers when executing a search query:

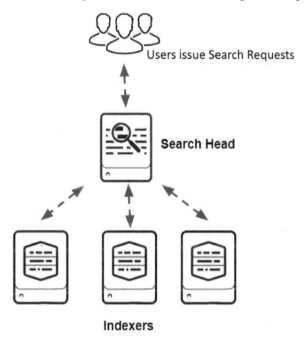

Figure 1.2: SH and indexers interaction

Let us look at another critical processing component—the indexer, which is also called a search peer, as it responds to queries issued by the SH.

Indexer

The indexer accepts and stores the indexed data, which can be retrieved later when requested by the SH. The sources of data transmission can include forwarder agents or inputs without requiring dedicated agents. The indexer(s) can be set up as either standalone instances or as a clustered configuration for HA. The data that has been indexed remains unchangeable and is stored in the form of buckets. More details about buckets are provided in *Chapter 5, Splunk Index Management*:

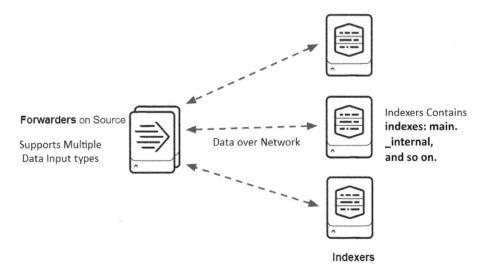

Figure 1.3: Indexers receiving data from forwarders and storing it in indexes

So far, we have gone through the processing components and their roles in a Splunk Enterprise deployment. Let us go through the management components in the following section.

Management components

These are management components that support the processing components:

- **Deployment Server (DS)**
- **SHC Deployer (SHC-D)**
- Indexer CM
- **License Manager (LM)**
- MC

We'll discuss them in the following subsections.

DS

A standalone Splunk Enterprise instance is used to manage the forwarders. The forwarders, which are located at the data source (typically a UF), often need new configurations to monitor new files or changes to an existing configuration followed by an optional restart. Changing them manually is a very time-consuming task in larger infrastructures. That's where the DS comes to the rescue, by maintaining a central repository of configurations in the form of apps. In addition to UFs, HFs can also be centrally managed using a DS.

Chapter 4, *Splunk Forwarder Management*, goes through more details on this topic.

SHC-D

The SHC-D manages app configurations and deployments for an SHC in Splunk Enterprise deployment. It distributes app bundles to the SHs, applies configurations, and coordinates rolling restarts if needed.

The SHC-D usually stores all the apps at the following location: `$SPLUNK_HOME/etc/shcluster/apps`.

Indexer CM

An indexer cluster incorporates a distinct Splunk Enterprise instance that functions as a Cluster manager, known as a CM. This CM does not engage in typical search operations but rather oversees the indexer cluster, governing it in the following ways:

- The **Replication Factor (RF)** is met
- The **Search Factor (SF)** is met
- Deployment of configurations to the cluster
- Responds to SH requests

The *Search head indexer clustering overview* section of *Chapter 7* will explain the RF and SF in detail.

License manager

All components in Splunk Enterprise require a license for commercial use, except for UF, which is a software offered by Splunk that is available for use without requiring a separate license. The LM is loaded with the license file received from Splunk sales by an admin. Multiple license files might exist depending on the agreement with Splunk. The rest of the instances in the deployment, called license peers, are connected to the manager node. The manager node acts as a central license repository for configuring stacks, pools, and license volumes. It stores usage logs in a `license_usage.log` file, which tracks all Splunk instances connected to the LM for violations and their usage. Out-of-the-box license reports are dependent on this log. We will discuss this in detail in *Chapter 2*, *Splunk License Management*.

Monitoring Console

The MC is a built-in app in Splunk that provides a centralized location for monitoring and managing Splunk deployments. It offers a GUI that allows administrators to monitor and configure various aspects of Splunk, including alerts and dashboards for monitoring indexing, license usage, search, resource usage, forwarders, health checks, and more. We will go through some of these dashboards in detail and set up alerts in later chapters.

> **Note**
>
> Do note that although these components have dedicated roles and activities to perform, some of them can be installed together on the same Splunk instance. A matrix of which components can be combined is provided in the docs: `https://tinyurl.com/26f9n5zf`.

We have come to the end of the components section. We learned that a UF is preferred for file monitoring and forwarding data to indexers. Depending on the deployment type, whether standalone or distributed, the number of components required to set up differs. Standalone Splunk doesn't require many components as it functions as both an SH and indexer. A distributed deployment includes a number of additional management components for deployment, cluster management, and license management. The Splunk Enterprise binary utilized for all components remains same; the differentiation lies in the configuration of each binary instance, determining the role of each component such as the SH, indexer, SHC-D, DS, or LM.

As we dive into the chapters associated with both processing and management components, we will look into these topics in more detail, and you will find them discussed a lot throughout the book. So, understanding these components and their role in Splunk Enterprise deployment is quite important to understand the rest of the sections and chapters.

Splunk Validated Architectures (SVAs)

This section is completely optional as this topic isn't included in the Splunk admin exam blueprint; however, I recommend going through it to get an insight and familiarize yourself with what Splunk's architecture looks like, as well as where the processing and management components are positioned and interconnected.

So far, we have learned about Splunk Enterprise's features and components and their roles in a standalone or distributed deployment. It is time to see some of the deployment architectures, called SVAs, curated by the best minds at Splunk Inc.

Just as there is more than one solution to a problem, similarly, a single architecture might not fit every organization. For Splunk Enterprise architects and Splunk Enterprise admins who go through many variables and evaluate to come up with a suitable design, SVAs offer guidance with best practices and off-the-shelf readily available designs. A Splunk Enterprise architect's roles and responsibilities vary from that of a typical admin. *Splunk Education* offers courses to prepare you to become a Splunk Enterprise-certified architect, and the Splunk Enterprise Admin certification is a prerequisite.

Let's go through some of the prominent validated architectures of Splunk Enterprise on-premises. A full list of SVAs is available here: `https://www.splunk.com/pdfs/technical-briefs/splunk-validated-architectures.pdf`.

Single-server deployment

A single-server alias standalone deployment consists of a Splunk Enterprise instance that combines both SH and indexer functionality.

The following diagram shows the deployment architecture:

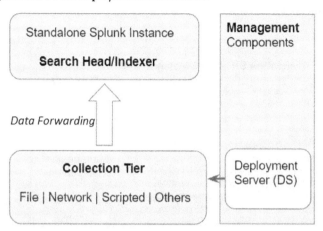

Figure 1.4: Standalone deployment architecture

The diagram shows a standalone/single Splunk instance, a collection tier forwarding events to a single instance, and an optional DS to manage the collection tier/forwarders.

The only advantage of this deployment type is its cost-effective and easy to manage.

Let's look at the limitations of this deployment type, as follows:

- Works for a limited number of users, between four and eight
- Data indexing size is limited to below 300 GB per day
- Does not work effectively for critical searches
- No high availability and disaster recovery
- Migrating to distributed deployment is straightforward with additional hardware

Let's take a look at distributed non-cluster deployment, which is a more advanced setup than a single-server deployment.

Distributed non-clustered deployment

Distributed non-clustered deployment works better for additional workload and indexing capacity than a single-server deployment. The separation of SH and indexing duties increases the **total cost of ownership** (**TCO**).

The following diagram shows the non-clustered deployment architecture with separate SH and indexing tiers:

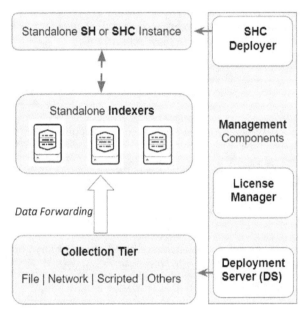

Figure 1.5: Distributed non-clustered architecture

In the depicted architecture, a separate search tier comprises a SHC and an indexing tier with multiple standalone indexers. The SHC-D is a mandatory management component responsible for deploying configurations to the SHC using apps. It facilitates the deployment process by pushing configuration updates via apps from the SHC-D to the SHC. A DS is utilized for managing forwarders, while an LM stores license information. The DS ensures effective forwarder management, while the LM serves as a central repository for license details, with all other instances connecting to it for license information. Let's look at the advantages of this deployment over single-server deployment:

- The number of users would be higher than in a single-server deployment with additional indexer support

- Independent indexers increase the daily indexing capacity to over 300 GB

Now, let's look at the limitations of this deployment:

- No HA and DR
- The SH needs to be reconfigured every time a new search peer/indexer is added
- Search results might be incomplete when one of the indexers is down, and data ingestion might be impacted as well
- In the case of a standalone single-SH deployment scenario, there is a **single point of failure** (SPOF)

Let's take a look at distributed cluster deployment, which is a more advanced setup than a distributed non-cluster deployment.

Distributed cluster deployment and SHC – single-site

A distributed clustered SH and indexer deployment at a single site is a highly available, resilient architecture. A site is a classic data center in a particular region/geography.

The following diagram shows the clustered deployment architecture with a separate SHC and clustered indexing tier running on a single site:

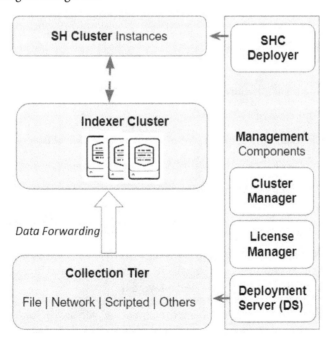

Figure 1.6: Distributed clustered deployment and SHC – single-site

Figure 1.6 shares similarities with *Figure 1.5*, as it depicts a similar architecture. However, in *Figure 1.6*, an additional management component, known as the CM, is introduced. The CM is responsible for overseeing and managing the indexer cluster, providing coordination and control of the cluster's operation. It acts as a central point for configuring and monitoring the indexers within the cluster, ensuring their effective functioning and synchronization. Let's understand the advantages of this over the two architectures we previously looked at:

- Using an SHC avoids an outage in the case of node failure by replicating the configs and artifacts. Job scheduling is managed by the captain, which is one of the elected nodes from the cluster itself.

- The indexer cluster enables data HA by maintaining redundant copies across the cluster. The CM is a separate management component that ensures data availability for searches.

- The SHC scales by adding more nodes compared to a single instance, allowing for increased capacity in handling concurrent users and executing searches. Typically, each CPU is considered as one unit for search capacity calculation, meaning one CPU counts as one search. This allows for better scalability and improved performance in handling larger workloads and user demands.

- Additional management components are the SHC-D and CM to aid in the deployment of apps to cluster members.

- The Indexer Discovery feature aids the SHC to discover when a new search peer/indexer is added. The SHC doesn't require reconfiguration for every new indexer node.

Now, let's look at its limitations compared to the previous architectures:

- No DR with a single site. A failure of a site will eventuate the failure of the entire deployment.

- The SHC has a limitation of 100 nodes.

- It increases the TCO and the management of the SHC and index cluster.

Let's take a look at the multi-site distributed clustered deployment, which is a more advanced setup than distributed clustered deployment and single-site.

Distributed clustered deployment and SHC – multi-site

This is by far the most complex architecture valid for organizations that have strict HA and DR requirements. It has the same advantages as single-site architecture (as seen in the previous section), and the failure of a site doesn't impact the entire deployment.

The following diagram shows a clustered deployment architecture with an SHC and clustered indexing tiers deployed in more than one site:

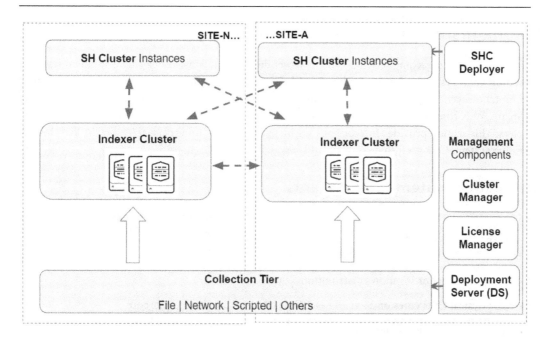

Figure 1.7: Distributed clustered deployment and SHC – multi-site

As in *Figure 1.6*, the components remain the same in each site. However, the collection tier is common across both sites. Each site has a dedicated SHC.

Let's understand the limitations:

- Indexer clusters replicate data between sites, which is called cross-site replication, requiring lower network latency. 100 milliseconds or less is preferred.
- SHCs work independently and do not share artifacts and common configurations.
- SHCs have a 100-node limitation per site.
- A dedicated SHC-D is required for each site.
- A single CM node suffices for an entire cluster of indexers across sites.

We've looked at a very basic single-server architecture (preferably used for testing or development) and an advanced multi-site cluster deployment architecture. Each has its advantages, limitations, and cost implications. At this stage, you are pretty much familiar with Splunk components and architectures. In the next section, we are going to install a standalone/single-server deployment, which we talked about at the very beginning of this section.

Splunk installation – standalone

As discussed in the preceding section, a single-server deployment consists of a single Splunk instance combining both SH and indexer functionality. The installation actually isn't part of the admin exam blueprint; however, it is very helpful to get your hands dirty by experiencing Splunk yourself through the Splunk Web, configuration file (`.conf`), and CLI options that we are going to discuss in upcoming chapters. This section provides instructions for installing Splunk Enterprise 9.0.3 on the Windows operating system. Let's get into it.

Installation system requirements

Let's look at the system requirements of the computing environment. Splunk Enterprise supports multiple operating system environments. A full list of the supported options is available here: `https://tinyurl.com/2tuudjwr`. Splunk has the following hardware requirements:

- A 64-bit Linux or Windows distribution

- 12 physical CPU cores or 24 vCPU @ 2 GHz or more clock speed per core

- 12 GB **random-access memory** (**RAM**)

- An x86 64-bit chip architecture

- 1 GB Ethernet **network interface card** (**NIC**)

- Free disk space of at least 3 GB for installation and more as per indexing needs

My system specifications for where Splunk version 9.0.3 is going to be installed are as follows:

- 64-bit Windows 11 Pro operating system

- 6 physical CPU cores (or 12 vCPUs) @ 2.1 GHz clock speed and 16 GB RAM

- An x86 64-bit AMD chip

- Plenty of disk space

You might have noticed the physical CPU cores in my PC are fewer than recommended, which is absolutely fine as we are not going to run production workloads on the Splunk instance. Let's get into the installation steps, as follows.

Installation steps

As a prerequisite, you need a high-speed internet connection to download the Splunk Enterprise free software package from here: `https://www.splunk.com/en_us/download.html`. If you do not have a Splunk account, then sign up and log in to continue. Choose the installation package by operating system and download the latest version, which is 9.0.3 at the time of writing.

Let's begin the installation:

1. Download the `.msi` file that appears as `splunk-9.0.3-dd0128b1f8cd-x64-release.msi`. Double-click on it to start the installation. You will be prompted to accept the license with the default installation options. Refer to *Figure 1.8* and click the **Next** button:

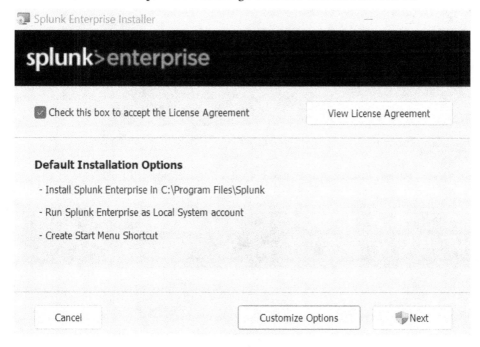

Figure 1.8: Installation – license agreement

2. You will be prompted to enter administrator account credentials. Enter the details. Make sure you remember them as you will need them to log in to the Splunk instance for the first time. Click the **Next** button (refer to *Figure 1.9*):

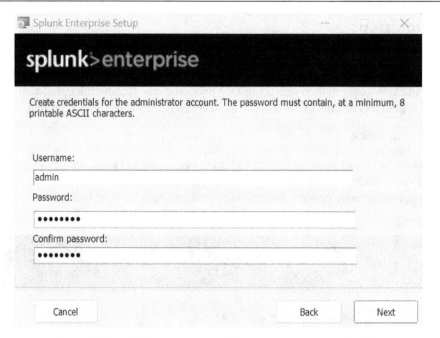

Figure 1.9: Installation – creating administrator account credentials

3. On the next screen, just click the **Install** button (refer to *Figure 1.10*):

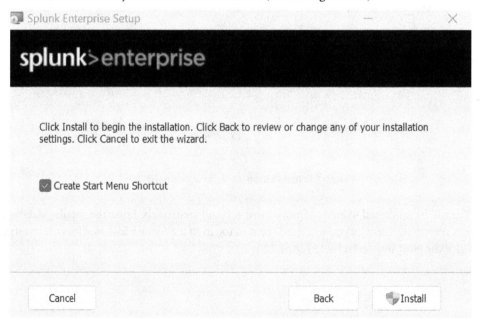

Figure 1.10: Installation – click Install to begin

4. The setup wizard takes a few minutes to install Splunk Enterprise. If all goes well, a final "successfully installed" screen appears, as shown in *Figure 1.11*. Clicking on the **Finish** button will launch the browser window:

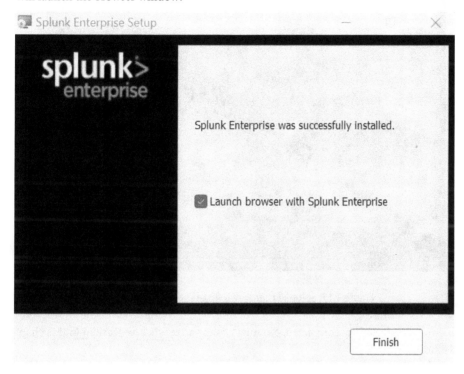

Figure 1.11: Installation successful

5. You should observe the first-time login browser window URL: `https://127.0.0.1:8000`. Here, `8000` is the default Splunk Web port and `127.0.0.1` is the loopback address. Enter the admin credentials created in *step 2*; then you will be taken to the Splunk Enterprise home page at `http://127.0.0.1:8000/en-GB/app/launcher/home`:

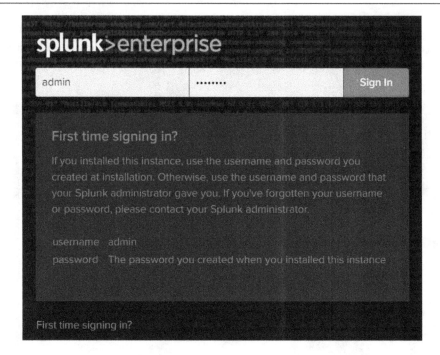

Figure 1.12: Splunk Enterprise – first-time sign-in page

The installation is successfully completed. Now, let's summarize what we learned in this chapter in the next section.

Summary

We have come to the end of the first chapter. There has definitely been a lot to digest. Let's briefly summarize what we have learned so far.

In this chapter, we began by looking at the Splunk Certified Admin certification prerequisites, the exam topics, and their weightage. In line with the exam topics, this book is organized into two parts: Splunk Enterprise system administration and data administration. We also discussed the exam pattern, which includes single- and multiple-choice as well as true/false questions.

We looked at the fundamentals of what Splunk Enterprise does and its key highlights as a data analysis product. We then progressed to look at the Splunk Enterprise 9.x product family features, followed by components and their role in deployment.

We also looked at prominent SVAs. We covered single-server, distributed non-clustered, distributed clustered single-site, and distributed clustered multi-site architectures. We discussed their advantages and limitations, showcasing processing and management components. Finally, we successfully installed a Splunk Enterprise single instance on a Windows system.

This chapter is the foundation for the rest of the book. The Splunk components that we looked at will be detailed in further chapters. It is required to know in what context they would be used and how they help in overall Splunk deployment architecture. Though SVAs are not part of the exam guide, they are included in the book to give you a better understanding of the upcoming chapters.

In the next chapter, we are going to deep-dive into license management. License management includes types of licenses, how they work, and license configuration.

In the next section, you are going to practice exam-style questions covering the topics that we have learned so far.

Self-assessment

This self-assessment section is to help you better understand which sections you are good at and which need improvements out of the topics covered in the chapter. I would suggest carefully reading the questions and answers and taking your time to go back through the sections that you think need more understanding. Alternatively, you could refer to the Splunk documentation. Good luck!

You will be given 10 questions and answers to choose from. The question patterns are the same as discussed in the *Introducing the exam's test pattern* section. At the end of this section, answers to the questions are provided. Let's get started:

1. Which of the following are Splunk Enterprise features? (Select all that apply)

    A. Alerting

    B. Search and reporting

    C. File monitoring

    D. Update files that are monitored as completed at the end of the file

2. Does Splunk support the monitoring of binary files?

    A. Yes

    B. No

3. The UF and Splunk Enterprise utilize which concept to prevent redundant data indexing?

    A. Bucket ID

    B. Fishbucket

    C. Finbucket

    D. HashBucket

4. What role does a SH play in a Splunk deployment? (Select all that apply)

    A. Contains dashboards

    B. Consolidates results from indexers

    C. Issues search queries to indexers

    D. Stores knowledge objects such as event types and macros

5. Indexers store data and respond to search queries. Is this statement true or false?.

    A. True

    B. False

6. Once data is stored in Splunk indexers, it cannot be directly modified by administrators, even if they wish to make changes. Is this statement true or false?

    A. True

    B. False

7. Which of the following are management components? (Select all that apply)

    A. UF

    B. **Search Head Cluster Deployer (SHC-D)**

    C. Cluster manager

    D. Deployment server

8. In a search head cluster environment, which among the following components is essential in a deployment?

    A. Monitoring console

    B. Cluster manager

    C. SHC-D

    D. Heavy forwarder

9. What does the **License Manager** (**LM**) contain? (Select all that apply)

    A. License file

    B. License stacks and pools

    C. License binary software

    D. License reports

10. Which configuration (`.conf`) file contains file monitoring details on a forwarder?

    A. `outputs.conf`

    B. `inputs.conf`

    C. `server.conf`

    D. `source.conf`

I hope you were able to recollect the topics that we went through with these questions. Let's review the answers.

Reviewing answers

1. *Options A, B, and C are correct answers. Option D is not a Splunk feature.*

2. *Option B, No.* Splunk cannot monitor binary files. It is only able to monitor text data.

3. *Option B, Fishbucket,* is the right answer. The Splunk Enterprise and the UF use seek pointers and CRCs to track the progress of file monitoring and avoid duplicate data indexing.

4. *Options A, B, C, and D* all are functions of an SH.

5. *Option A is correct (True).* Indexers store raw events, index files, and other metadata files for search processing.

6. *Option A is correct (True).* Once the data is written to an index, it is immutable.

7. *Options B, C, and D* are the correct answers. The UF is a data collection component.

8. *Option C is the right answer.* The SHC-D deploys apps to all cluster members through the captain.

9. *Options A, B, and D* are the correct answers. The LM stores the license file upon adding a license. Optionally, licenses are stackable, and pools can be created on the LM. License reports are available in the MC app, available in the LM instance.

10. *Option B is the correct answer.* `outputs.conf` and `server.conf` are Splunk configuration files used for different purposes. There is no `source.conf` file available in Splunk.

Splunk License Management

One of the key responsibilities of a Splunk system administrator is license management, When you install Splunk Enterprise for the first time, either for personal or commercial use, you need to configure the license to enable its functionality; otherwise, its default license validity will expire in 60 days. By the end of this chapter, you should understand license types, installation, how Splunk licensing works, license allocation, and license violations, and we will conclude with the license usage report in the Monitoring Console.

The overall goal of this chapter is to familiarize you with Splunk licensing so that you will be able to choose the right license type, stack your license file if you have more than one type, and configure and allocate the right volumes to peers. In this chapter, we will cover the following topics:

- Introducing license types
- Understanding license warnings and violations
- How licensing works
- Installing, managing, and monitoring licenses

Introducing license types

Before directly jumping into the subject of types, let's discuss the need for licenses. Splunk Enterprise is paid proprietary software that provides services to organizations that need a data platform that aligns with their business objectives. Splunk sells these services in the form of licenses to organizations to activate the Splunk Enterprise product. If you need help choosing the correct license for your use case, you can contact the Splunk sales team (sales@splunk.com).

We will discuss the types of Splunk licenses in the following sections.

The Splunk Enterprise Trial license

The Trial license is enabled by default when you first install Splunk Enterprise. The Trial license has the following features:

- It is valid only for 60 days. After this period, other license types must be selected to continue using the product.

- It allows you to use the full features of Splunk Enterprise.

- It works only for single/standalone Splunk Enterprise instances.

- The license limits data ingestion to 500 MB per day. Beyond this, Splunk will raise a license warning. License violation is explained in detail in the *Understanding license warnings and violations* section.

The Splunk Free license

The Free license is valid forever and supports only a subset of Splunk Enterprise features:

- It disables alerting, authentication, clustering, distributed search, summaries, and forwarding to non-Splunk servers.

- The daily data ingestion limit is 500 MB. License violations block the search functionality. More details are provided in the next section, *Understanding license warnings and violations.*

- It works only for single/standalone Splunk instances.

The Forwarder license

Splunk offers a variety of forwarders. The most popular ones are the **universal forwarder** (**UF**) and **heavy forwarder** (**HF**). Let's see how this license applies to the two main types of forwarders:

- The UF is by default embedded with the license, so there is no need for explicit license installation/configuration.

- The UF is separate binary software, and it consumes fewer resources than the HF. The UF is able to forward data to indexers with authentication.

- The HF is a full Splunk Enterprise instance binary that requires either a Forwarder license or an Enterprise license.

- The Forwarder license is by default embedded into the Splunk Enterprise package.

- No indexing is allowed on HFs with the Forwarder license. However, authentication is allowed.

- HFs that require indexing, advanced authentication, and the KV Store need the Enterprise license.

The Splunk Enterprise license

The Splunk Enterprise license must be purchased from Splunk. This license limits the per-day indexed data capacity. It offers a comprehensive suite of enterprise functionalities, including an indexer, deployment server, search head, deployer, and cluster manager. Now, let's examine the features associated with the Enterprise license:

- The search functionality in an Enterprise license stack with a data volume of 100 GB per day or more does not currently experience any disruptions. A stack below 100 GB per day that is in violation will be blocked from searching after 45 warnings in a rolling 60-day period.

- It permits standalone to complex distributed deployment architectures.

- More than one enterprise license can be stacked and allocated to license pools.

The Splunk Enterprise license is volume-based, which means it calculates usage using the daily indexed data size. Splunk has introduced a new license type that is based on the infrastructure size – that is, the number of vCPUs. Let's take a look at it in the next section.

The Splunk Enterprise infrastructure license

The Splunk Enterprise infrastructure license must be purchased from Splunk with reference to the number of vCPUs. It allows the full set of enterprise features. Let's look at the features of this license:

- It does not enforce license violations if the number of vCPUs exceeds the license; this might change in future versions, so always refer to the Splunk docs for the latest updates.

- It permits standalone to complex distributed deployment architectures.

- More than one enterprise infrastructure license can be stacked and allocated to license pools.

- The Splunk Enterprise infrastructure license cannot be stacked with the Splunk Enterprise license.

Splunk Developer license

As the name suggests, this license is for Splunk app developers to utilize Splunk Enterprise to develop and test apps. It enables the full set of Splunk Enterprise features. Here are the features of this license:

- It allows 10 GB a day of data. Exceeding the data size will raise license warnings. Searching will be blocked if the number of warnings exceeds the limit. Refer to the next section, *Understanding license warnings and violations*, for more details.

- It is issued for a 6-month validity period. A renewal request can be raised to extend the license.

- It cannot be stacked with any other licenses.

In conclusion, Splunk offers a variety of licenses, from the starter's Trial license to the commercial Enterprise and infrastructure licenses. The Free license only offers a subset of features and never expires.

The Developer license is purely for app developers and enables all the Splunk Enterprise features. Every license enforces limits – for example, licenses are limited to a number of days, data volume, vCPU count, and so on. The following section elaborates on the generation of license warnings, how they transition to violations, and subsequently details the specific features that Splunk Enterprise restricts for each license type.

Understanding license warnings and violations

Understanding license limitations and their impact on the Splunk Enterprise platform is very important if system administrators want to maintain the continuity of their platform. The license types, whether purchased from Splunk or free, enforce limits, except the infrastructure license, which counts the number of vCPUs and doesn't enforce license violation limitations. The rest of the license types enforce limits on data volume or indexed data volume per day.

If a license exceeds the allowed data volume per day, Splunk Enterprise will raise a license warning. When the number of warnings exceeds a certain threshold, it is called a license violation. The thresholds differ for every license type. A license violation disables the important features of Splunk Enterprise.

License warnings appear in Splunk web administrative messages. They persist for a number of days in order for system administrators to resolve them. License usage alerts can be enabled in the monitoring console.

Figure 2.1: Monitoring console alerts – license usage and expiry

Figure 2.1 shows the alerts you can receive in the monitoring console. They are disabled by default; you can enable them and discover how they work in the event that usage is close to the daily quota and/or your license is near expiry.

License violation disables important Splunk features, which means that only a subset of features will be available during the violation period. Let's see what happens to search and indexing in a violation period:

- Search functionality, which includes reports and alerts, will be blocked, except for internal indexes such as `_internal`, `_audit`, and `_introspection`
- Indexing of data will continue for indexers

Let's discuss the threshold for warnings and violations for each license type:

- **Splunk Enterprise Trial license**: Five or more warnings in a rolling 30-day period is a violation. License warnings persist for 14 days. During the violation period, searching is blocked and indexing continues. If you wish to fix the license violation to restore the search functionality, there is no reset option. The suggested approach is to reinstall the whole Splunk Enterprise instance.

- **Splunk Free license**: Three or more warnings in a rolling 30-day period is a violation. License warnings persist for 14 days. During the violation period, searching is blocked and indexing continues. Similarly to the Trial license, the Free license cannot be reset, and reinstallation is the only option.

- **Splunk Enterprise license**: A license stack with 100 GB or greater data volume per day does not enforce license warnings and violations. A license stack with less than 100-GB data volume per day does raise license warnings. 45 or more warnings in a rolling 60-day period is a violation, and searching is blocked while indexing continues. To reset the Enterprise license, contact the Splunk sales team; it will be able to unlock the required search features.

- **Splunk Enterprise infrastructure license**: No violations are raised for this license type.

- **Splunk Developer license**: Five or more warnings in a rolling 30-day period is a violation. License warnings persist for 14 days. During the violation period, searching is blocked and indexing continues. The Developer license can be reset to unlock the search features by contacting the Splunk developer program team at `devinfo@splunk.com`.

- **Splunk Forwarder license**: No violation rules are applicable to the Forwarder license. It is an embedded license in the Splunk Enterprise package. Indexing on the Forwarder license requires either an Enterprise or infrastructure license.

In conclusion, Splunk license warnings tell administrators upfront how the overall license usage is so that they can take the required action before a violation occurs. A sequence of warnings exceeding the defined threshold over a period of time will lead to a violation. During the violation period, searching gets blocked except for internal indexes, and indexing continues. Neither the Enterprise stack license over 100 GB nor the Enterprise infrastructure license will enforce violations, but the rest of the license types enforce violation rules. A violated license can be reset, except for the Free and Trial ones.

We have covered a lot of information about license warnings and violations; however, you might have wondered at what stage Splunk Enterprise tracks license usage, and which component tracks the usage in distributed deployment architectures. The following two sections will focus on license installation in a license manager component, and we will discuss in detail how exactly licensing works with regard to license usage calculations.

How licensing works

Splunk Enterprise licensing tracks the indexed data volume per day against a license stack or individual pools. To deep-dive into license groups, stacks, and pools, refer to the Splunk documentation: `https://docs.splunk.com/Documentation/Splunk/9.0.3/Admin/`

Groups, stacks, pools, andotherterminology. The contents of this documentation are covered in the next section, *Installing, managing, and monitoring licenses.*

In a distributed infrastructure, a dedicated Splunk Enterprise instance will be assigned a license manager role, which keeps track of the license usage of the whole deployment. The rest of the instances act as license peers reporting to the manager.

The license quota is measured every day, from midnight to midnight, which means it will get reset to the full quota for the next day's processing, and it's a continuous process that works in conjunction with the system clock of the manager. Splunk supports two types of indexes, called **events indexes** and **metrics indexes**. They are covered in *Chapter 5, Splunk Index Management.* License usage calculations are capped to a fixed 150 bytes per metric event for metrics indexes and, for event indexes, the full data size of the events.

The original data from the source will go through various phases before getting indexed to indexers. The data size is measured at the indexing phase. Phases are explained in *Chapter 8, Getting Data In.* So far, we have talked about data being indexed from external sources. There is a heap of data that is generated locally on every running Splunk Enterprise instance. It is stored in internal indexes that start with _ (underscore), namely _internal, _audit, _introspection, _metrics, and _telemetry.

Let's see which data doesn't count in the license quota:

- Splunk internal logs don't count

- Replicated copies are stored in the indexer cluster for high availability

- Summary indexes with the stash source type

- Indexed data, as it contains tsidx files, metadata, and so on to enable faster retrieval

In conclusion, license quotas are measured every day, from midnight to midnight. The data size is calculated during the indexing phase. Metrics index data is capped at 150 bytes per event, and event index data is calculated based on the full size of the events. Internal logs, summary indexes, and replicated data don't get measured against the license. A dedicated instance works as the license manager, in which the license file is installed and configured; the remaining instances function as license peers and report to the license manager in a distributed deployment.

Installing, managing, and monitoring licenses

Now that you have a comprehensive understanding of license types and how license usage is calculated against the license quota, we will discuss adding, configuring, and managing a license. As a system administrator, these tasks are quite important. We will take the example of the Splunk Enterprise license type to discuss the following topics:

- Adding a license

- License stacking and pools

- License manager and license peer

- License usage and alerting

Let's discuss each of these in detail in the following sections.

Adding a license

A Splunk Enterprise license is a file with the `.license` extension, which is generally purchased from Splunk. The following steps are applicable to other types of license, such as the Developer license.

A license file can be added to Splunk in either of the following two approaches:

- On the Splunk web home page (`https://<hostname>:8000`), go to the menu in the top-right corner of the window and navigate to **Settings | Licensing**. Click on the **Add license** button, and on the next screen, select **Choose File** to upload a `.license` file obtained from Splunk sales. Then, click the **Install** button. At this stage, you are required to restart the Splunk instance. The following screenshot shows the licensing web page, and you can see the **Add license** button, which is used to upload a license file.

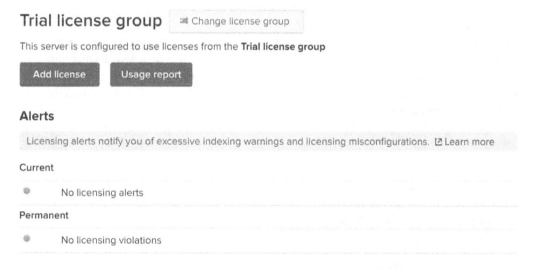

Figure 2.2: The Licensing web page

- The second approach is to execute a CLI command on the host the Splunk instance runs on. You should have access and execution permissions to the host where Splunk license manager is installed. A restart is required after the successful command execution:

```
splunk add licenses <path_to_license_file_required>
```

In both approaches, Splunk Enterprise stores the license file in the `$SPLUNK_HOME/etc/licenses` directory. The following screenshot shows the location of the license files added to the Splunk instance running on Windows:

```
C:\PROGRAM FILES\SPLUNK\ETC\LICENSES
+---download-trial
|       enttrial.lic
|
\---enterprise
        Splunk (1).License.lic
        Splunk (2).License.lic
```

Figure 2.3: The license files' location

In the following section, let's take a look at scenarios in which more than one license has been added to the Splunk instance and how they get grouped and stacked together.

License groups, stacks, and pools

The installation of a valid license file enables all the required features of Splunk Enterprise. However, in certain scenarios, it may be necessary for more than one license to be installed. It's common to implement the license allocation strategy to effectively manage different deployment environments. For instance, when dealing with a development environment alongside a production environment, it's essential to prevent the development indexers from consuming the entire license volume, thereby leaving adequate capacity for the production workload. This is one of the scenarios in which system administrators take advantage of the advanced management topics of license groups, stacks, and pools.

License groups

License groups contain a set of license stacks. There are four license groups, and only one group can be active at a time on the license manager in a distributed deployment. To understand this in more depth, let's go through the group details:

- **Enterprise Trial group**: By default, this is active when Splunk is first installed. More than one Trial license cannot be stacked.

- **Enterprise free group**: This is the next best group after the Enterprise trial expires. More than one free license cannot be stacked.

- **Forwarder group**: This is dedicated to forwarders that don't do indexing. Forwarder licenses cannot be stacked.

- **Enterprise license/sales group**: This contains an Enterprise license purchased from Splunk or a sales license received from Splunk for product evaluation. They can be stacked, which means an Enterprise license of 50 GB and a Sales license of 10 GB provisions 60 GB of data per day.

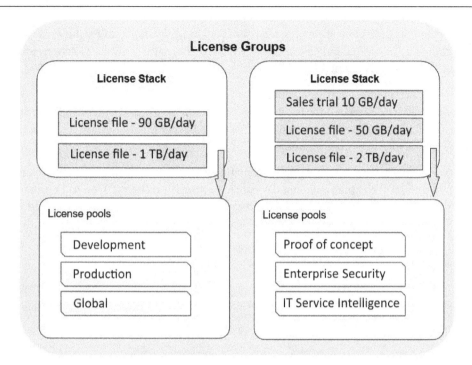

Figure 2.4: The license group, stack, and pool layering

License stacks

License stacks contain one or more related licenses whose license quotas can be added together to contribute toward the permitted data volume per day.

For example, an initial purchase of a 50-GB Enterprise license can be stacked with the later purchase of a 40-GB license, which together allows an indexing volume of 90 GB per day. Let's look at the key points to remember when stacking licenses:

- Enterprise licenses and sales licenses can be stacked together
- A license that is already part of one stack cannot be part of another stack
- License warnings and violations will be raised against stacks and pools

License pool

A license pool is a portion of the license volume of a stack that can be assigned to different indexers. It allows a granular level of control over the allocation of licenses.

For example, a total of 100 GB of license needs to be allocated for both development and production indexers. However, the development environment could use the whole volume just because it is allocated. During the development process, a burst of data indexing might utilize the whole license, leaving nothing for the production environment, which might lead to a license violation, which in turn impacts production environment.

To ring-fence a portion of the license, a pool will be created and allocated to development indexers. Distinct pools will be designated for production, proof of concept, and other premium Splunk applications such as Enterprise Security and IT Service Intelligence (ITSI). Refer to the license pools portion in *Figure 2.4* for more information.

The following figure shows the license pool information created for the demonstration. To create a pool, open the Splunk Web home page and navigate to **Settings | Licensing**. Then, click on the **Add pool** button and follow the instructions. I have created two pools – one called `dev-indexers`, which is allocated 5,240 MB, and another, `auto_generated_pool_enterprise`, with 5,000 MB, which is for non-development indexer use.

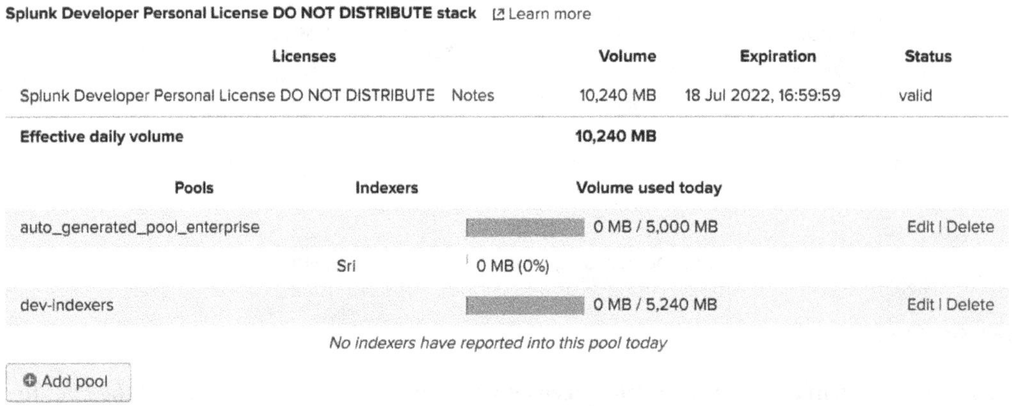

Splunk Developer Personal License DO NOT DISTRIBUTE stack   ⬚ Learn more

Licenses	Volume	Expiration	Status
Splunk Developer Personal License DO NOT DISTRIBUTE Notes	10,240 MB	18 Jul 2022, 16:59:59	valid

Effective daily volume			**10,240 MB**	

Pools	Indexers	Volume used today		
auto_generated_pool_enterprise		0 MB / 5,000 MB		Edit I Delete
	Sri	0 MB (0%)		
dev-indexers		0 MB / 5,240 MB		Edit I Delete

No indexers have reported into this pool today

⊕ Add pool

Figure 2.5: License stacks and pool allocation

License manager and license peers

In a standalone Splunk Enterprise environment, a single Splunk Enterprise instance works as a search head and indexer and maintains its own license, so there is no need for a license manager. In a distributed deployment, duties are separated across search heads, indexers, HFs, and optional management components.

A license manager is a Splunk Enterprise instance that is configured to act as a centralized license server, and the rest of the instances in the deployment report to this server. The remaining instances in deployment that are configured to report to the manager instance are called license peers. The license stacks, pools, and their volume allocation are configured on the manager instance.

In the *Adding a license* section, we learned about installing a license on a Splunk instance that acts as a repository of licenses, which is called a license manager. Log in to the license manager in which licenses have been installed, go to the top-right corner of the window, and then go to **Settings | Licensing**. Refer to *Figure 2.5* to see the licensing web page, which contains information on the licenses, volumes, expiration, and license pools. In a distributed deployment, other Splunk instances are called license peers. All the peers need to point to the license manager instance to obtain license information.

To configure a license peer, log in to the remaining instances in the distributed deployment, such as search heads, indexers, and forwarders, and point them to the license manager. Log into the individual instance web interface and navigate through **Settings | Licensing**, as shown in *Figure 2.2*, and click on the **Change to Peer** button, which shows two radio button options, **Designate this Splunk instance <instance_name> as the manager license server** and **Designate a different Splunk instance as the manager license server**. As shown in *Figure 2.6*, choose the second option, which will display a text box for you to enter the manager license server URI and its management port (8089). Enter manager license server URI and click the **Save** button to connect the license peer to the license manager.

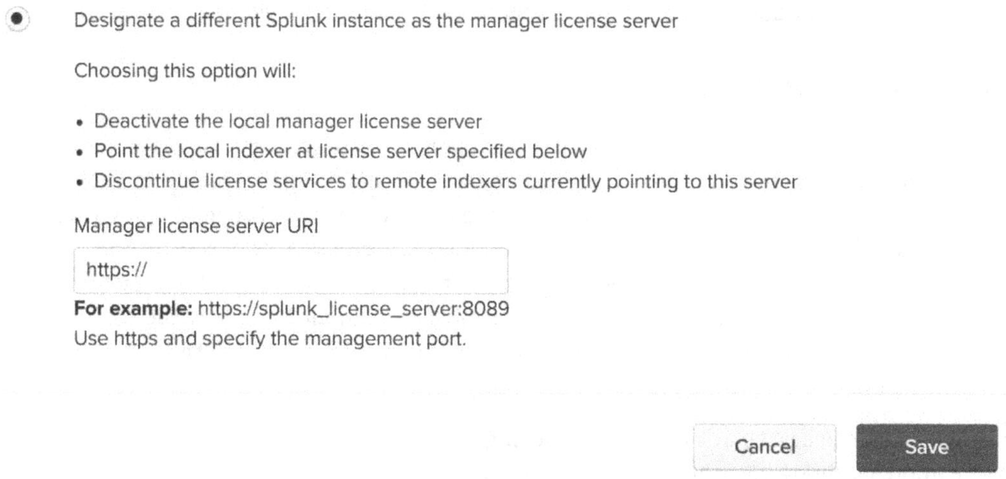

◉   Designate a different Splunk instance as the manager license server

    Choosing this option will:

    • Deactivate the local manager license server
    • Point the local indexer at license server specified below
    • Discontinue license services to remote indexers currently pointing to this server

    Manager license server URI

    `https://`

    For example: https://splunk_license_server:8089
    Use https and specify the management port.

                                                       Cancel   Save

Figure 2.6: Configuring a license peer with the license manager URI

The following figure shows the **Messages** section, where license warnings and violation messages appear in Splunk Web. License warnings and violation messages appear on the license manager's Splunk Web home page in the **Messages** menu bar.

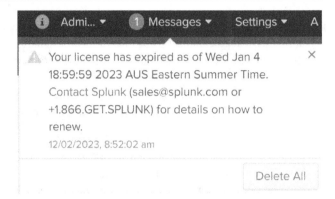

Figure 2.7: License warnings in Splunk Web

In the event that a license peer is unable to reach the license manager, there will be a grace time of 72 hours before the peer is declared to be in license violation. In this scenario, the search functionality will be blocked, but indexing continues until the peer restores connectivity with the manager instance.

License usage and alerting

The monitoring console is included by default in Splunk Enterprise software. However, it must be configured to use in either standalone or distributed mode. Configuring the monitoring console is out of scope of this book, however, if you would like to explore the topic, I suggest that you follow the Splunk documentation (`https://tinyurl.com/msxtcmkj`). To access the monitoring console application, log into the Standalone Splunk instance web interface and navigate through the **Settings** menu and click on the **Monitoring Console** option to the left side of the menu items. It provides two visual dashboards, which can be found in the **Indexing | License Usage** menu of the monitoring console.

Figure 2.8 shows two dashboards under the **Indexing** menu. The **License Usage - Today** dashboard shows comprehensive details about today's license usage. To check past usage, open the **Historic License Usage** dashboard.

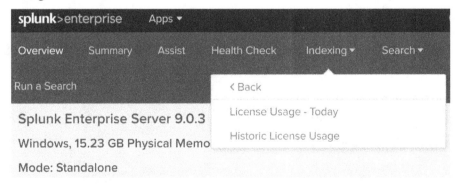

Figure 2.8: License usage dashboards in the monitoring console

Similarly, the monitoring console contains two alerts for license usage tracking. Refer to *Figure 2.1* for alert details. Alerts are disabled by default. To monitor license usage information, alerts must be enabled manually so that they will send notifications when the license expires in 2 weeks' time, and if the daily license quota has exceeded 90%. The notification message appears on the Splunk Web home page in the **Messages** menu bar.

There is one more place where warning and violation alerts are recorded on the Splunk Web home page – under the **Settings | Licensing** menu (refer to *Figure 2.2*). The alerts raised here will have a more granular level of visibility about the pools and stacks that are in a current or permanent licensing violation.

In conclusion, a license can be installed either through Splunk Web, the Splunk CLI, or config files. A license manager is a repository of all licenses, and the rest of the Splunk instances report to the manager and are called license peers. If a license peer is unable to report for 72 hours, it will be declared in license violation, and its search feature will be blocked (although indexing continues) until it reconnects to the license manager instance.

License groups, stacks, and pools help to manage licenses effectively if more than one license is installed, and they allocate a portion of the volume to specific indexers. Only one license group will be active at any point in time. Only certain types of licenses are allowed to stack – for example, an Enterprise license and a Sales license can be stacked, but an infrastructure license cannot be stacked with the Enterprise license. The monitoring console provides license usage reports and alerts to track the usage. Alerts are disabled by default. License warnings and violation messages appear in Splunk Web in two places – one is on the Splunk Web home page in the **Messages** bar, and the other is under **Settings | Licensing**.

After the *Summary* section, there are some practice exam questions covering the topics that we have learned so far.

Summary

We have come to the end of the license management chapter, which has been very comprehensive. Licensing is native to Splunk Enterprise, and the naming conventions take a bit of time to digest.

We began with the licensing types that Splunk provides and then went through the differences between them to examine the suitability of different licenses for different deployment options. Splunk provides Enterprise Free and Trial licenses and Developer licenses with some limitations. To enable all the features of Splunk Enterprise, you need a valid license file. If a license is used beyond its limitations, a license warning will be raised, and the license will be treated as being in violation after a certain number of warnings. There are no violations for the infrastructure license, and the Enterprise license has a quota of 100 GB or more. License violations block important features, such as searching, but indexing continues. The Free and Trial licenses cannot be reset from violation, but the others can be reset by reaching out to Splunk Sales.

Afterward, we went through how license works and is calculated in the indexing phase. There are exclusions, such as summaries, replicated copies, and internal logs, which will not count toward the licensing volume.

Finally, we dived deep into how to install, manage, and monitor licenses. We provided screenshots explaining how to configure them via the web and the CLI. In distributed deployments, a license manager is a repository of licenses, and the rest of the instances act as license peers. We saw how to switch between a license manager and a license peer. We also saw what happens if a license peer disconnects from the manager for 72 hours – the peer will be declared in violation. During the peer's violation period, the search feature will be blocked (although the indexing feature continues) until the peer reconnects to the manager instance. Finally, we looked at the built-in monitoring console's dashboard locations and useful alerts to monitor license usage.

In the next chapter, we'll discuss user and authentication management in Splunk.

Self-assessment

Here, you will test yourself on the concepts that you have covered in this chapter. You will be given 10 questions and answers on license management; at the end of this section, the answers are provided. The pattern of the questions is the same as we discussed in the *Introducing the exam's test pattern* section in *Chapter 1*. Don't be alarmed if you struggle with certain concepts, as they involve quite new terminology and take a bit of time to understand. Always refer to the relevant sections of the chapter when you feel you have difficulty answering the questions; alternatively, you can refer to the Splunk documentation. Good luck! Let's get started:

1. For how many days is the Splunk Enterprise Trial license valid?

    A. 55 days

    B. 60 days

    C. 50 days

    D. 90 days

2. What is the Splunk Free license's daily ingestion limit?

    A. 1,024 MB/day

    B. 512 MB/day

    C. 500 MB/day

    D. 100 MB/day

3. Which of the following statements is true about the Splunk infrastructure license? (Select all that apply):

    A. It is measured against the number of vCPUs.

    B. The ingestion volume is limited to 100 GB/day.

C.   License violations never occur.

D.   License violations stop indexing.

4.   Which Splunk components require a license in a distributed deployment and need to connect to the license manager? (Select all that apply):

A.   Universal forwarder

B.   Search head

C.   Heavy forwarder

D.   Indexer

5.   Which indexes are excluded from the license calculation? (Select all that apply):

A.   `main`

B.   `_internal`

C.   `_audit`

D.   `_introspection`

6.   Does Splunk enforce installing a separate license for the metrics index?

A.   Yes

B.   No

7.   What is the cap for the metrics index to calculate the license usage?

A.   15 bytes

B.   150 bytes

C.   500 bytes

D.   100 bytes

8.   A company called irobot has a Splunk Enterprise license of 200 GB per day. Will there be a license violation if a 200-GB ingestion rate is exceeded in a day?

A.   Yes – three warnings in a rolling 30-day period.

B.   Yes – 45 warnings in a rolling 60-day period.

C.   No, there won't be a license violation for a license over 100 GB.

D.   Yes – 10 warnings in a rolling 30-day period.

9.   Is a license pool a portion of volume created from a license stack?

A.   Yes

B.   No

10. What is the CLI command used to add licenses in Splunk, and usually in what location is the license file stored?

   A. Command: `./splunk add license-file <path_to_license_file>`, Location: $SPLUNK_HOME/etc/licenses

   B. Command: `./splunk install license-file <path_to_license_file>`, Location: $PLUNK_HOME/etc/license-stack

   C. Command: `./splunk add licenses <path_to_license_file>` Location: $SPLUNK_HOME/etc/licenses

   D. Command: `./splunk add licenses <path_to_license_file>` Location: $SPLUNK_HOME/etc/license-groups/license-stacks

I hope you were able to recall the licensing topics when going through these questions. Let's review the answers.

Reviewing answers

1. *Option B* is the correct answer. The Splunk Enterprise Trial license is valid for 60 days from the date you install it; this license is applied by default to your instance, which you can change at **Settings | Licensing**.

2. *Option B* is the correct answer. The Splunk Free license limits the data volume to 500 MB/day. Exceeding the volume and generating three or more warnings in a rolling 30-day period will be a violation. During violation, searching is blocked and indexing continues.

3. *Options A and C* are the correct answers. There is no enforcement of license violation for infrastructure licenses, and per-day volume limits are related to the Splunk Enterprise license.

4. *Options B, C, and D* are the correct answers. The universal forwarder contains an embedded license, so it doesn't need to be connected to the license manager. The remaining components must connect to the license manager in a distributed deployment.

5. *Options B, C, and D* are the correct answers. All Splunk internal indexes (which start with `_<index_name>`) are excluded from license calculations. Similarly, summary indexes with the stash sourcetype don't count toward the license quota.

6. *No*, metrics indexes don't require a separate license. Metrics events are capped at 150 bytes per event to calculate the license usage.

7. *Option B*: 150 bytes is the correct answer.

8. *Option C* is correct. Enterprise licenses over 100 GB don't enforce license violation.

9. *Yes*, license pools can be created from license stacks.

10. *Option C* is the correct answer.

Users, Roles, and Authentication in Splunk

Splunk Enterprise follows the **role-based access control** (**RBAC**) approach, which allows users to access Splunk instances but restricts what users see and which actions they can perform. As a Splunk system administrator, it's important to familiarize yourself with Splunk's default user roles as well as to learn how to create custom roles to meet your organization's specific requirements. You may also be tested on your understanding of role inheritance, which allows you to assign multiple roles to a user, and how to manage index access for users. It's important to show that you can set up authentication in Splunk using methods like LDAP, SAML, and more. These skills are tested in the Splunk Enterprise Admin certification exam, as they are crucial for effectively managing user permissions and access in a Splunk environment.

The following are the key components involved in taking the RBAC approach, which we will discuss throughout this chapter, along with how to configure them through Splunk Web and the Splunk CLI:

- Users
- Roles
- Authentication methods

Users

A Splunk Enterprise instance, as discussed in previous chapters, offers a variety of interfaces—namely, Splunk Web, the Splunk CLI, and RESTful APIs. All the interfaces must be secured by allowing users to log in securely through authentication and authorization. The authentication methods are explained at the very end of the chapter. Organizations can effectively manage user authorization by configuring roles, capabilities, object-level permissions, and **Role-Based Access Control** (**RBAC**) to align with individual users' job responsibilities. Users are then able to perform administrative tasks (privileged users) or general user tasks (the creation of reports, alerts, dashboards, and so on) depending on the Splunk role assigned to them. A user must be assigned at least one role in Splunk. The following screenshot shows user management menu items on the Splunk Web home page, under **Settings**:

USERS AND AUTHENTICATION

Roles

Users

Tokens

Password Management

Authentication Methods

Figure 3.1: Splunk users and authentication management

The following page is shown upon clicking the **Users** menu item from *Figure 3.1*. The **Users** page contains a list of existing users' full details with an **Edit** action and a **New User** button to create users:

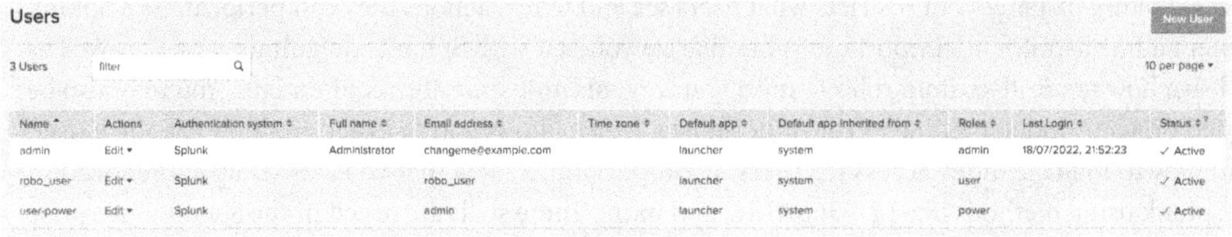

Name	Actions	Authentication system	Full name	Email address	Time zone	Default app	Default app inherited from	Roles	Last Login	Status
admin	Edit ▾	Splunk	Administrator	changeme@example.com		launcher	system	admin	18/07/2022, 21:52:23	✓ Active
robo_user	Edit ▾	Splunk		robo_user		launcher	system	user		✓ Active
user-power	Edit ▾	Splunk		admin		launcher	system	power		✓ Active

Figure 3.2: Splunk user management page

For example, the administrator will create new users and allow them to search a particular index by assigning a role. The user created by the administrator will be able to log in and run a search against the allowed index.

Users are entities who perform required actions on the Splunk platform in a secure way. Without any users, a Splunk instance is just orphaned and isn't used for its purpose. For the same reason, you will be prompted to create an administrator user account after the installation the first time you boot a Splunk instance.

Let's add a new user through Splunk Web and find out which details are required in the following section.

Creating a new user

A new user can be created either through Splunk Web or the Splunk CLI. They both require administrator account details to authenticate a Splunk instance. System administrators should be aware of at least these details to complete the user creation process: the user's full name and the roles they can be assigned or the capabilities they need to have on the Splunk platform.

Let's look at user creation through Splunk Web. First, log in to Splunk (`https://<hostname>:8000`) using administrator credentials and navigate to **Settings**, then click on the **Users** hyperlink to find the **New User** button. Clicking on the button will open a **Create User** form, as shown in *Figure 3.3*:

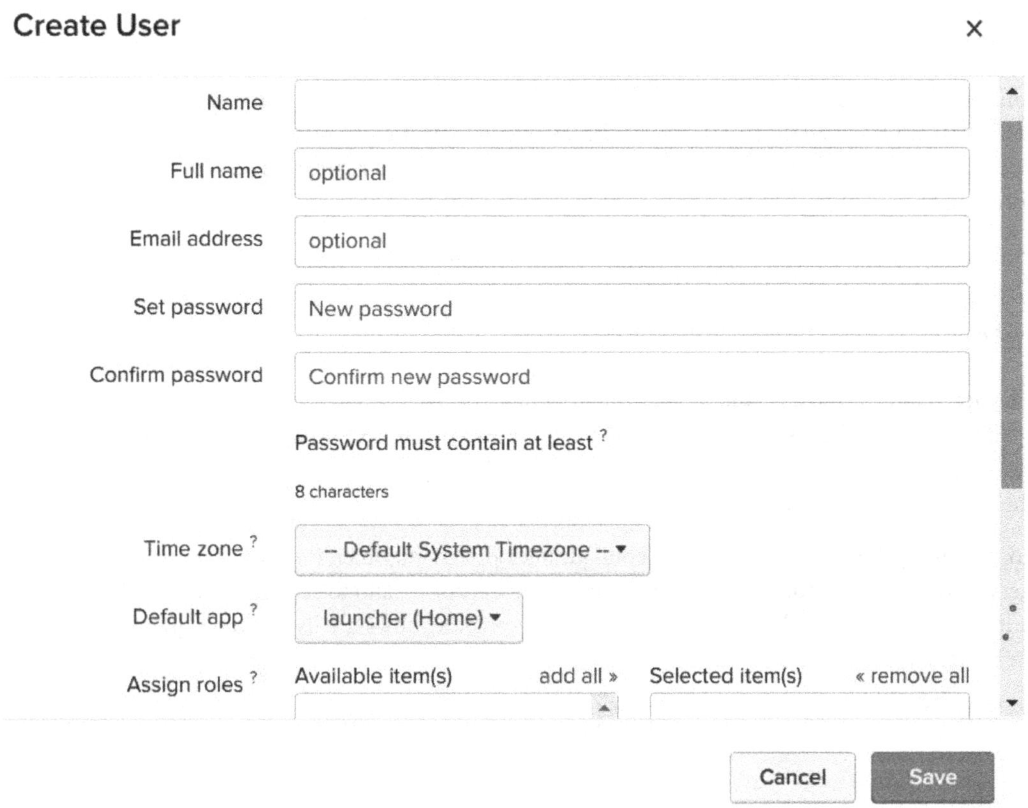

Create User   ✕

Name	
Full name	optional
Email address	optional
Set password	New password
Confirm password	Confirm new password
	Password must contain at least ?
	8 characters
Time zone ?	-- Default System Timezone -- ▾
Default app ?	launcher (Home) ▾
Assign roles ?	Available item(s) add all » Selected item(s) « remove all

Cancel   Save

Figure 3.3: Creating a Splunk user

The **Name** value in the form should not contain any spaces, and it will be the username entered when a new user logs in to a Splunk instance.

Complete the **Create User** form by filling in the name, full name, and password, and assign at least one role. There is a checkbox at the end of the form to force the user to change the password on the first login, which is not visible in *Figure 3.3*; however, it can be enabled if required for privacy, so the user will be able to choose their own password.

Let's take a look at the CLI command to add a user called `supportuser` with the password `changeme2`, assigning the `supportgroup` role, which is a custom role. Here, `admin:changeme` is the administrator account username and password separated by the `:` (colon) symbol:

```
./splunk add user supportuser -password changeme2 -role supportgroup
-auth admin:changeme
```

Similarly, there are more CLI commands available to edit the existing user, unlock the user, reset the user's password, and so on. You can refer to the Splunk documentation at `https://docs.splunk.com/Documentation/Splunk/9.0.0/Security/ConfigureuserswiththeCLI` for more details.

Let's look at the roles and their association with users in the following section.

Roles

Roles are assigned to users, and at least one role must be assigned to a user. Roles contain a list of capabilities that defines which actions users can perform on the Splunk platform. In other words, capabilities are nothing but a list of permissions given to a role—for example, run a search, run a scheduled search, run a real-time search, and so on.

Along with capabilities, a role provides configurable options for which indexes the user is allowed to query, to restrict them further through a search filter approach, and to set quotas such as disk-space limit and user-level search limits. The following screenshot shows the list of configurable options a role contains:

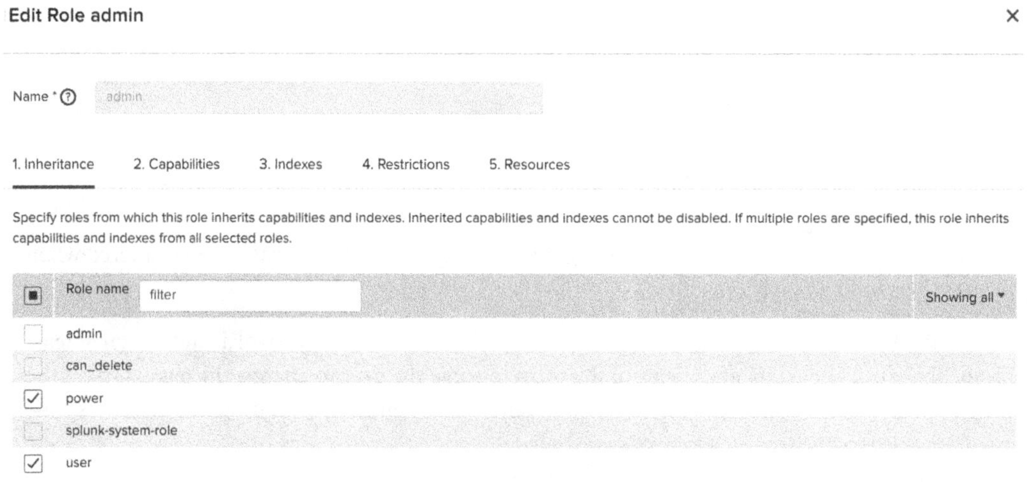

Figure 3.4: Edit Role configurable options

Splunk Enterprise ships with the following default roles. Along with them, a Splunk administrator can create a custom role by inheriting an existing role. Custom roles that inherit existing roles inherit the parent role capabilities along with allowed indexes, filters, and resources.

You might be wondering why there is a need to define new custom roles when default roles already exist in Splunk. The reason is they provide flexibility, scalability, and the **segregation of duties (SoD)**, allowing organizations to effectively manage user access and permissions, aligning with their specific needs, and ensuring proper data security and compliance.

The following are the different roles, and their details, that a Splunk system administrator needs to understand and manage as part of their responsibilities:

- `user`—Allows the user to create/edit/delete their own saved searches, dashboards, reports, tags, event types, and so on.

- `power`—This inherits the user role. In addition, the user can share their own objects with other users, such as alerts, dashboards, tags, and so on, and edit objects shared by other users.

- `admin`—This inherits both the `user` and `power` user roles and has the most capabilities. The user can create other admin users and custom roles, modify system settings, and so on.

- `can_delete`—This role allows the user to delete already indexed events with the `delete` command. Once deleted, data cannot be reverted. A `delete` marker will be applied to events. You should be cautious about assigning this role to a user.

Roles can be accessed by logging in to the Splunk Web home page, navigating to the menu bar in the top-right corner, and selecting the **Settings | User and Authentication | Roles** menu item, as shown in *Figure 3.5*:

Roles

New Role

5 Roles   filter

Name ▲	Actions	Native capabilities	Inherited capabilities	Default App ⇅
admin	Edit ▾	106	32	
can_delete	Edit	4	0	
power	View Capabilities	9	23	
splunk-system-role	View Indexes	0	138	
user	Clone	23	0	

Figure 3.5: Role management page

Creating a new role

I hope you now have a comprehensive understanding of roles and their significance and are aware of the default Splunk roles. In this section, let's find out about the need to create a custom role and go through the role creation steps.

For example, to understand the requirement for new custom roles, take two different indexes: the `web-access-data` index contains website access log data, and the `building-access-data` index contains building access log data. The website access log data will only be accessible to application support teams, and the building access log data will only be accessible to the building security team. In this scenario, there will be two different roles created for the two teams.

Default Splunk roles can be assigned to users; however, it depends on which indexes they need access to, which capabilities should be allowed, and the allocation of resources such as disk quota, search quota, and so on. If an existing role doesn't suit the needs of the user, the systems administrator can create a new role by optionally inheriting an existing Splunk role. A new role inheriting an existing Splunk role inherits the parent role capabilities, index access, and resource limits.

Splunk doesn't recommend changing default roles. If you want to edit the default roles, the suggested approach is to clone the existing role and change the required configurable options. The following screenshot shows the **New Role** creation form:

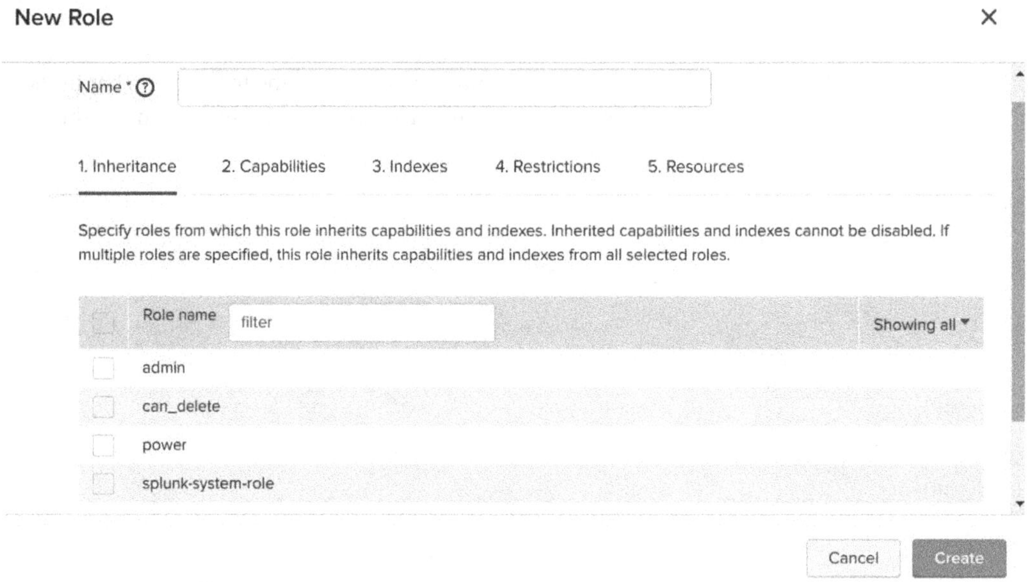

Figure 3.6: Creating a new Splunk role

The preceding screenshot shows the name to be given to the role, and five tabs: the **Inheritance** tab to choose an existing role, the **Capabilities** tab to choose the capabilities of the role, the **Indexes** tab to choose the indexes accessible by the role, the **Restrictions** tab to set a search filter on indexed fields, and finally, the **Resources** tab to limit the disk space and the search quota allowed at the role level and user level. Clicking the **Create** button will create a new role, which will appear on the **Roles** web page, as shown in *Figure 3.5*.

Capabilities are granular-level permissions applied to users, and the list is quite big. You can refer to the documentation at https://tinyurl.com/bdffj2s6 for a full list. Another place to look at them is the authorize.conf **specification (spec)** at https://tinyurl.com/2p8nvat9, which contains an up-to-date list of capabilities as they are subject to change.

So far, you have seen the process of creating/editing an existing role via Splunk Web. Splunk Enterprise, in general, contains configuration files (.conf) on the filesystem for every setting that appears in Splunk Web. Splunk writes every change that's made in Splunk Web to a configuration file. Let's take a look at the authorize.conf file, which contains role configuration details for every tab that appears in *Figure 3.6*.

Let's take a look at a sample authorize.conf snippet for the admin role. role_admin is called a stanza in Splunk terminology, and the format of the stanza is role_<name_of_role>:

```
[role_admin]
# ==== Subsumed roles ====
importRoles = power;user
# ==== Capabilities    ====
accelerate_datamodel    = enabled
admin_all_objects       = enabled
### Many such capabilities enabled
# ==== Other settings ====
srchIndexesAllowed = *;_*
srchIndexesDefault = main;os
srchFilter    = *
srchDiskQuota    = 10000
srchJobsQuota    = 50
```

The authorize.conf file, by default, exists in the $SPLUNK_HOME/etc/system/default directory. If you want to create a new role through the configuration approach, Splunk recommends creating a local directory either in the $SPLUNK_HOME/etc/system/local system directory or in the $SPLUNK_HOME/etc/apps/<app-name>/local app local directory. To effect the changes being made to configuration files, a restart of Splunk instances is recommended.

For full specs of all the Splunk configuration files, go to the $SPLUNK_HOME/etc/system/README directory. You will find files with .spec and .example extensions. That is all about roles. I advise you to practice with them by creating them through Splunk Web and the authorize.conf spec file to familiarize yourself with the terminology. Navigate through each of the tabs shown in *Figure 3.6*.

Let's look at the authentication methods supported by Splunk in the following section.

Authentication methods

Authentication methods are the ways a user is validated when an attempt is made to log in to a Splunk Enterprise instance. Splunk could either deny or allow access to the respective user. Successfully authenticated users will be authorized to access resources and perform actions according to the roles assigned to the user. A user with no role can't log in to a Splunk instance, so at least one role must be assigned to the user at the time of user creation.

Splunk supports multiple authentication methods, which is beneficial for organizations due to various reasons. Firstly, compliance requirements may mandate the implementation of **multi-factor authentication** (**MFA**) to enhance security, and Splunk provides support for MFA to meet these requirements. Secondly, internal employees of organizations can utilize the **single sign-on** (**SSO**) feature offered by Splunk, enabling them to log in without having to enter separate user credentials. This streamlines the authentication process and improves the user experience. Thirdly, some organizations prefer to leverage their existing **Active Directory** (**AD**) for authentication purposes, and Splunk allows integration with AD for seamless authentication. Lastly, in situations where other authentication methods such as an **identity provider** (**IdP**) for SSO or the **Lightweight Directory Access Protocol** (**LDAP**) are not reachable, Splunk provides the option of using a local admin account for login and troubleshooting purposes, ensuring access and availability even in challenging circumstances.

Let's go through the authentication methods and what they offer:

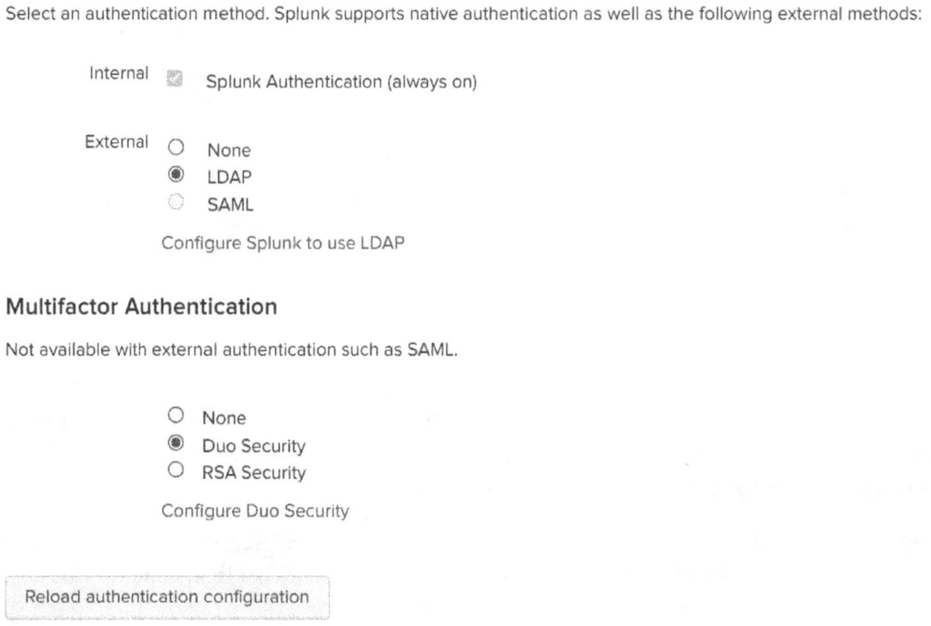

Select an authentication method. Splunk supports native authentication as well as the following external methods:

Internal ☑ Splunk Authentication (always on)

External ○ None
⦿ LDAP
○ SAML

Configure Splunk to use LDAP

Multifactor Authentication

Not available with external authentication such as SAML.

○ None
⦿ Duo Security
○ RSA Security

Configure Duo Security

Reload authentication configuration

Figure 3.7: Authentication method selection

The preceding screenshot shows the internal authentication (native Splunk), external authentication, and MFA methods. If you select the required radio button, that will display a hyperlink to configure the respective method. For example, if you choose the **LDAP** radio button, it enables **Configure Splunk to use LDAP**. Click on that link to complete the configuration of LDAP. Similarly, you could select other methods. You need to be aware of the LDAP server details and/or **Security Assertion Markup Language (SAML)** IdP details to complete the next steps.

Native Splunk

The native authentication method is enabled by default when Splunk is installed. It is always available to use regardless of any other external authentication method configuration. Native user credentials in encrypted format are stored locally on the Splunk Enterprise instance in the `SPLUNK_HOME/etc/passwd` file.

A blank `passwd` file disables native authentication. The best practice is to have at least one native admin account actively available to use in the event an external authentication method is unavailable. The Splunk Web URL can be forced to use a native method by appending a query parameter as follows:

```
https://<hostname>:<port-default-8000>/en-GB/account/
login?loginType=splunk
```

LDAP

LDAP is an external authentication method able to connect to a variety of directory services, such as Microsoft AD, OpenLDAP, and so on.

As shown in *Figure 3.7*, the **LDAP** option is available by selecting the radio button. Upon choosing it, a **Configure Splunk to use LDAP** hyperlink is available to click and configure it, as shown in the screenshot. Then, click on the **New LDAP** button and fill in the required LDAP strategy details so that for every login attempt by the user, Splunk will be able to retrieve the LDAP groups of the respective user. After the creation of an LDAP strategy, on the same page, click **Map groups** in the **Actions** column for a specific strategy.

LDAP groups are the named entries assigned to an employee or user of a particular organization. The Splunk administrator maps the LDAP groups to Splunk roles.

There are at least 30 steps involved in LDAP strategy creation and map groups combined. You could go through the Splunk docs and follow the steps to set up LDAP authentication: `https://tinyurl.com/37wxs9ue`.

The one noticeable difference between LDAP/SAML and native Splunk authentication methods is the users created in a Splunk Enterprise instance, by default, are assigned the native authentication method. It is not possible to assign users manually to either LDAP or SAML authentication methods.

SAML

SAML is useful for enabling an SSO authentication scheme on the Splunk platform so that users will be able to log in through SSO identification. The Splunk platform supports the SAML 2.0 protocol version and contacts the IdP to retrieve the user's SAML group information. Similar to LDAP, the SAML groups are mapped upfront to Splunk roles by Splunk administrators.

There is very good documentation available from Splunk to configure SSO using SAML-based authentication. You can refer to the docs here: `https://tinyurl.com/294zrc8u`.

MFA

MFA is supported by the Splunk platform through Duo and RSA security mechanisms, as shown in *Figure 3.7*. This method will let users authenticate to the Splunk platform using more than one service for enhanced security.

Configuration is done through the Splunk Web **Settings | Authentication methods | Multifactor Authentication** menu, selecting one of the radio buttons (**Duo** or **RSA**), then clicking on the **Configure Duo Security** hyperlink, filling in the form, and clicking the **Save** button.

If you want to deep-dive into the detailed steps, follow the Splunk documentation link: `https://tinyurl.com/sv8rnwhb`.

Scripted authentication

The Splunk platform provides a script-based authentication method for an external system such as **Remote Authentication Dial-In User Service (RADIUS)** or a **pluggable authentication module (PAM)**. There is no direct option in Splunk Web to configure it; however, the sample scripts are available under `SPLUNK_HOME/shared/authScriptSamples`, where you can follow the instructions in the README file.

Summary

In conclusion, roles, users, and authentication methods all together secure a Splunk Enterprise instance. We started by gaining an understanding of users. They are the primary entities that do the required actions and executions on the Splunk platform. Users can be either administrators (admins) or general users who will be created by the admin. We went through the creation of a new user through the Splunk Web interface and the Splunk CLI. Afterward, we looked into what a role has to offer for a user.

Splunk follows the RBAC approach. Roles are created and managed by system administrators. Splunk offers default roles, and they can be extended by inheriting them through the new role creation process.

Assigning a user to a specific role is a commonly used solution in Splunk to restrict their access to certain actions, limit their consumption of resources, and control their access to indexes. Roles in

Splunk allow for fine-grained permissions management, ensuring that users have appropriate privileges and restrictions based on their assigned roles.

A user must be assigned to at least one role and can be assigned to multiple roles if required. We went through new role creation via Splunk Web and the configuration file approach. A config file change requires a Splunk instance reboot.

Lastly, we looked at supported authentication methods such as native Splunk, LDAP, SAML 2.0, **two-factor authentication** (**2FA**), and script-based authentication methods. Native Splunk is an always-on authentication method, and the remaining methods require integration with external authentication systems such as a directory server for LDAP, an IdP for SAML, DUO/RSA for 2FA, and RADIUS or PAM for scripted authentication.

In the next chapter, we are going to look into the collection layer, which is nothing but forwarders. You will learn about forwarder management through the deployment server, installation, and configuration, and monitor them using the MC application.

Self-assessment

The self-assessment section is to test yourself on what you have gone through in this chapter. You will be given 10 questions and answers to choose from. The question pattern is the same as what we discussed in the *Introducing the exam's test pattern* section in *Chapter 1*. After the assessment, refer to the sections related to the questions you are having difficulty answering. Alternatively, you could refer to the Splunk documentation. All the best! Let's get started:

1. A Splunk user requires at least one role to be assigned when created. Is this statement true or false?

    A. True

    B. False

2. Which roles, by default, exist in Splunk after installation? (Choose all that apply)

    A. `splunk`

    B. `admin`

    C. `user`

    D. `can_delete`

3. What are the Splunk CLI functions to manage a user? (Choose all that apply)

    A. Reset password

    B. Unlock user

    C. Add user

    D. Edit user

4. You have been tasked with adding a new capability to the existing `power` user role. What will you do? (Choose all that apply)

    A. Edit the `power` role

    B. Clone the `power` role and add a new capability

    C. Create a new role similar to the `power` role and add a new capability

    D. Clone the `user` role and add a new capability

5. A Splunk Enterprise instance requires an admin user. Is this statement true or false?.

    A. True

    B. False

6. What are the authentication methods supported by Splunk? (Choose all that apply)

    A. LDAP

    B. SAML

    C. Splunk internal

    D. RADIUS

7. Your manager asked you to delete data in a specific index. Which actions will you take? (Choose all that apply)

    A. Use the `delete` command

    B. Log in as an admin user inherited from the `admin` role

    C. Log in as an admin user inherited from the `admin` and `can_delete` roles

    D. Log in as a power user

8. Which `.conf` file contains roles and capabilities information?

    A. `authentication.conf`

    B. `authorize.conf`

    C. `roles.conf`

    D. `capabilities.conf`

9. A new employee has joined the team and needs Splunk access. The Splunk instance is configured to LDAP authentication. What will you do as an administrator?

    A. Create a local user that will synchronize with LDAP after the first login

    B. Create an LDAP account for the employee

    C. Suggest the employee gets added to the LDAP group to which Splunk authenticates

    D. Run the Splunk CLI `./splunk add user` command to create a user account

10. How can you force Splunk to use native authentication when the SAML authentication method is configured on the Splunk instance?

    A. Open Splunk Web on the default `8000` port

    B. Open Splunk Web on port `8000` with the `param login=splunk` query

    C. Open Splunk Web on port `8000` with the `param loginType=splunk` query

    D. Open Splunk Web on port `8089`, which is reserved for native authentication

I hope you were able to recollect the topics that we went through about users, roles, and authentication. Let's review the answers in the next section.

Reviewing answers

1. *True*—you won't be able to create a user without a role.

2. *Options B, C, and D* are the correct answers. The default roles are `user`, `power`, `admin`, `can_delete`, and `splunk-system-role`.

3. *Options A, B, C, and D* are all correct answers.

4. *Options B and C* are the right answers. Splunk Inc. does not recommend editing the default roles. Instead, always clone default roles for modification or create custom roles. Inheritance of an existing role is a great feature as well.

5. *True*—every component in a Splunk instance, including the SH, indexer, HF, and management components, requires an administrative account to perform essential administrative tasks, such as configuration, monitoring, and management of the respective component. Admin accounts ensure the necessary privileges to carry out these administrative functions effectively.

6. *Options A, B, and C* are the correct answers. Splunk internal authentication is also called native Splunk authentication.

7. *Options C and A* are the right answers. An admin user should have the `can_delete` role to execute the `delete` command. Just be mindful about executing this command—you cannot undo it.

8. *Option B* is the correct answer. There are no such files as `roles.conf` or `capabilities.conf` in Splunk. `authentication.conf` contains settings related to authentication.

9. *Option C* is the correct answer. When adding a user locally on Splunk, they will not be authenticated against LDAP, and their credentials will not be synchronized with LDAP. To enable LDAP authentication and synchronization, the user needs to be assigned to the appropriate LDAP group which is mapped to the Splunk roles. This ensures that their authentication is performed against the LDAP directory, and any changes or updates to their account are synchronized with the LDAP server.

10. *Option C* is the correct answer. For example – `https://splunkhostname:8000/en-US/account/login?loginType=splunk`.

4

Splunk Forwarder Management

The *forwarder* name should sound familiar at this stage as you will have already read about two types of forwarder: the **universal forwarder** (**UF**) and the **heavy forwarder** (**HF**). These are two different software binaries built for specific use cases. In this chapter, you will learn more about UFs, which are managed centrally through the **deployment server** (**DS**) in large and complex Splunk Enterprise environments.

We will begin by learning more about the purpose of UFs in the overall Splunk deployment architecture, followed by configuring the DS, and the installation and configuration of UF. Configuration includes connecting the UF to external indexers for data forwarding and using the `deploymentclient.conf` configuration to download apps from the DS. Finally, you will be introduced to monitoring forwarders through the monitoring console application.

This chapter holds significant importance not only for system administrators but also for data administrators. It involves shared responsibilities between both types of administrators, including installing the UF, DS, and indexer connection configuration that are prerequisites for data administrators to create the necessary data inputs on the forwarder to ingest data into indexers.

In this chapter, we will cover the following topics:

- Introducing the universal forwarder
- Configuring the deployment server
- Installing the universal forwarder
- Configuring forwarding on the universal forwarder
- Configuring `deploymentclient`
- Forwarder monitoring

Introducing the universal forwarder

A UF is another software package from Splunk Inc. that is different from the Splunk Enterprise binary. A UF is installed and configured on the source systems where data originated and can collect the data from a range of sources and forward it to Splunk indexers for indexing. For example, to forward web access logs generated by a web server stored in a file to Splunk indexers, the UF is installed on the web server host with read permissions to the log files. The UF actively monitors the logs, securely transmitting the data to the designated Splunk indexers for indexing.

In the context of UFs and Splunk indexers, the data receivers that the UF connects to are essentially the indexers themselves. Indexers are configured to listen on a designated port (typically 9997 by default). Once the UF has been installed, it is directed to the IP/DNS address of the Splunk indexers and the corresponding receiving port to establish a connection. This allows the UF to securely send the collected data to the indexers.

Let's go through the key details of UFs:

- UF software is free to download and no license is required. As we discussed in *Chapter 2, Splunk License Management*, the license is embedded into the UF.

- The UF supports various **operating system (OS)** versions and CPU architectures. Choose the right one by reading through the instructions at `https://www.splunk.com/en_us/download/universal-forwarder.html`.

- The resource usage (memory, CPU, disk, and so on) of the UF is less than Splunk Enterprise.

- The UF does not have a UI interface, so configuration is typically done by directly editing configuration files, using Splunk **command-line interface (CLI)** commands, or utilizing a Splunk DS if the UF is configured to communicate with it.

- The UF cannot search, index, or parse data. However, there are exceptions for structured file formats such as CSV, W3C, TSV, PSV, and JSON. For these structured file formats, the UF can perform parsing by utilizing the `INDEXED_EXTRACTIONS` setting in the `inputs.conf` configuration file.

- The UF utilizes a concept called the **fishbucket** to prevent the same file contents from being forwarded to the Splunk indexers. In the event of a host restart where the UF is running, the UF resumes reading the file from where it left off, based on the last read location. The fishbucket is a mechanism used by the UF to track processed files. It stores metadata, including the last read position, so that when the UF restarts, it knows where to continue reading. This prevents duplication of indexed data and ensures that only new or updated data is forwarded to the indexers, ensuring data integrity and efficiency in file monitoring:

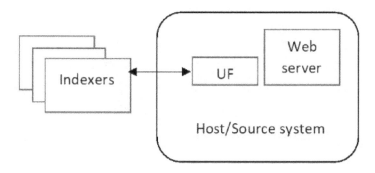

Figure 4.1: UF existence in a source host

In the preceding figure, the UF is installed on a host where the web server is running. The UF intends to read the web server logs and forward them to indexers.

Let's go through the configuration of the DS, followed by installing and configuring the UF.

Configuring the Deployment Server

The DS, as its name suggests, centrally manages the deployment to forwarders. It can deploy to both UFs and HFs and is optionally able to restart them after apps have been deployed. In this section, you will find out about the capabilities of DS and the configuration required to manage the forwarders.

Let's go through the DS setup requirements:

- The DS is an instance of Splunk Enterprise software that requires a Splunk Enterprise license.

- Forwarders, whether they're UFs or HFs, require a `deploymentclient.conf` file to be configured to establish a connection with the Splunk DS. This configuration file enables the forwarder to "phone home" and communicate with the DS. By either manually editing the `deploymentclient.conf` file or using the Splunk CLI, the forwarders can establish the required connection with the DS.

- The `serverclass.conf` configuration file for managing server classes can be set up through the Splunk Web interface by navigating to the **Settings | Forwarder Management** menu. The typical location of the `serverclass.conf` file on the Splunk DS is `$SPLUNK_HOME/etc/system/local`. Refer to *Figures 4.2* and *4.3* for more details about the Splunk Web interface.

- Splunk apps/add-ons can either be custom-built by following the Splunk app structure or downloaded from `splunkbase.com`, where hundreds of pre-built apps/add-ons are available.

- Apps/add-ons that need to be deployed to forwarders are placed in the `$SPLUNK_HOME/etc/deployment-apps` directory.

- Apps that are deployed through the DS are stored in the forwarder client's `$SPLUNK_HOME/etc/apps` directory.

Let's look at the configuration of `serverclass` in the next section.

Configuring serverclass

To configure the DS via the Splunk web forwarder management approach, at least one app must be copied to the `$SPLUNK_HOME/etc/deployment-apps` directory; you can download any sample app from `splunkbase.com`. Log in to `splunkbase.com` using your Splunk account to download the app.

Once you've copied the app to the `deployment-apps` directory, proceed with the following steps:

1. After copying the app, log in to Splunk Web as an administrator and navigate to **Settings | Forwarder Management**:

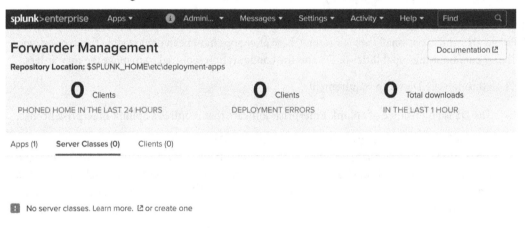

Figure 4.2: Splunk Web – Forwarder Management

2. *Figure 4.2* shows **Forwarder Management** and three tabs: **Apps**, **Server Classes**, and **Clients**. The first time you access this area, no server classes will exist (as shown by the zero after the tab's name), so click on the **create one** link.

3. The link will take you to a page where you can enter the class's name and choose the app and client to map the app for deployment. *Figure 4.3* shows two server classes, **AppsForLinux** and **AppsForWindows**, and the **Apps** count shows **1** for each:

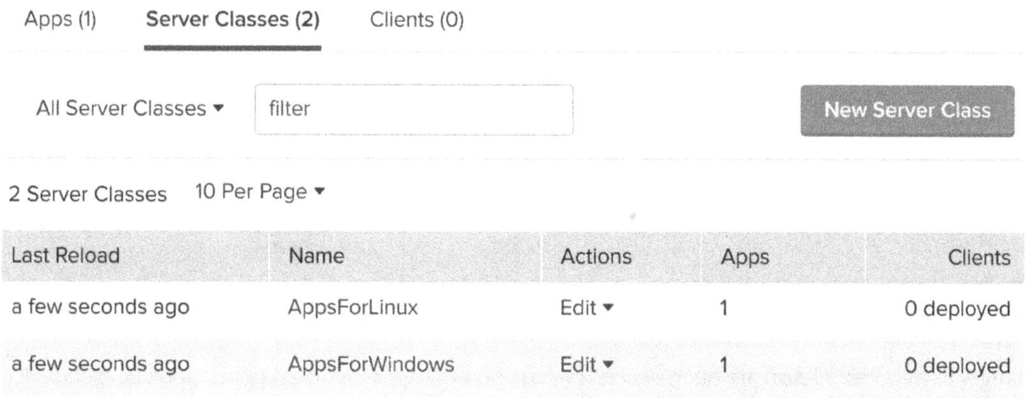

Apps (1)   **Server Classes (2)**   Clients (0)

All Server Classes ▾ | filter | [New Server Class]

2 Server Classes   10 Per Page ▾

Last Reload	Name	Actions	Apps	Clients
a few seconds ago	AppsForLinux	Edit ▾	1	0 deployed
a few seconds ago	AppsForWindows	Edit ▾	1	0 deployed

Figure 4.3: Splunk Web server class configuration

As shown in *Figure 4.3*, you can create as many server classes as you want by clicking on the **New Server Class** button. A server class to the client is a one-to-many mapping, which means that if you want to deploy a Splunk Linux app to all the UFs running in Linux OS environment across the organization, you can do this with a single server class.

The `serverclass` feature in Splunk provides two filtering options: `whitelist` (to match specific forwarder clients) and `blacklist` (to exclude specific clients). Clients can be configured based on their client name, hostname, IP address, and DNS name, or using **Perl Compatible Regular Expressions** (**PCRE**)-compliant regular expressions as patterns.

The configuration format for these settings follows an increased sequence, ordered as `whitelist.<n>` and `blacklist.<n>`, where n represents an unsigned integer. The sequence can begin at any value and is not necessarily consecutive.

Some apps require the forwarder client to be restarted after deployment. In that case, setting `restartSplunkd = true` in the `serverclass` file does the restart after deployment. The same can be enabled in the Splunk **Forwarder Management** UI while adding clients to the server class.

Let's look at the `serverclass.conf` file that's been saved on the filesystem:

```
# Deploying Linux App to all the hosts matching host dns #*.host.
yourcompany.com, web.yourcompany.com
# reloads after installation and restarts client if
# necessary
[serverClass:AppsForLinux]
whitelist.0= *.host.yourcompany.com
whitelist.1= web.yourcompany.com
issueReload=true
restartIfNeeded=true

# App 'Linux' mapping to serverClass defined
[serverClass:AppsForLinux:app:Splunk_TA_nix]
```

```
# Deploying Windows App to matching IP address 10.9.8.*
# and reloads client after installation
# and restarts if necessary
[serverClass:AppsForWindows]
whitelist.0=10.9.8.*
issueReload=true
restartIfNeeded=true

# App 'Windows_App' mapping to serverClass defined
[serverClass:AppsForWindows:app:Windows_App]
```

Changes to serverclass are often necessary, based on the apps that need to be added or excluded for specific categories of forwarder clients. When any new forwarder clients are deployed across the organization, central management might be needed, requiring adjustments to the forwarder's hostname, DNS, or IP address, along with the installation of new apps or add-ons for forwarders. These are some well-known scenarios and there are many such cases that could arise.

At this stage, the DS is ready to accept incoming connections from forwarders or deployment clients. In the following section, we'll install the UF and configure the deploymentclient.conf file.

Installing the universal forwarder

The installation steps for the UF are different for various OSs, and the UF software package differs too, depending on the OS and computing architecture. In this section, you will find only Windows and Linux-compatible UF installation instructions.

To find the list of UF software packages available by OS, go to https://www.splunk.com/en_us/download/universal-forwarder.html.

Before we dive into the direct installation, Splunk suggests checking the hardware prerequisites. The following CPU, memory, and disk requirements apply to UF 8.2.x through 9.x on all OSs:

- **CPU**: 1.5 GHz clock speed
- **Primary memory**: 512 MB
- **Disk**: 5 GB space

Let's take a look at the Windows installation instructions.

Installation in Windows OS

To demonstrate, I have downloaded the Windows 11 64-bit architecture-compatible .msi forwarder package from the preceding download link. The package can be installed through the command line/PowerShell, through an interactive **graphical user interface** (**GUI**) approach, or by silently installing the Windows machine if this is enabled with user account control.

You can find all the options at the following link, along with additional information. They are completely optional for the certification – it is just for your reference: `https://tinyurl.com/5n73nk7e`.

Let's go ahead and do the installation using the GUI:

1. To install the Splunk UF, you need to double-click on the `.msi` installation package file, which typically starts with `splunkforwarder`. For instance, the following file has a name that contains `9.0.1`, which indicates the downloaded version of the UF – that is, `splunkforwarder-9.0.1...x64-release.msi`.

2. Follow the on-screen instructions. Start by checking the box to accept the license agreement, which will enable the **Next** button. Leave the default options as-is.

3. Clicking on the **Next** button will take you through creating credentials for an administrator account. Enter the desired username and password details, and note them for later use. Click the **Next** button until you find the **Install** button.

4. The **Deployment Server** and **Receiving Indexer** details are optional and you can ignore the warning. Finally, click the **Install** button and wait for the **Installation completed** message. If Windows prompts for a restart, then do so.

5. The UF should automatically start. You can check this by going to **Windows Services** and searching for `Splunk`.

The default installation location in Windows is `C:\Program Files\SplunkUniversalForwarder`, which is also referred to via the `SPLUNK_HOME` environment variable. To check the UF's status, go to the preceding location's `$SPLUNK_HOME\bin` folder and execute the `splunk.exe status` command; you should see a `Running` status output. Refer to *Figure 4.4*:

```
 Administrator: Command Prompt


C:\Program Files\SplunkUniversalForwarder\bin>splunk.exe status
SplunkForwarder: Running (pid 5476)
```

Figure 4.4: Splunk UF status check

Now, let's look at how to install the UF in the Linux OS.

Installation in Linux OS

The hardware requirement specification for UF installation in the Linux OS is the same as the Windows OS. You must use CLI commands to install it on Linux. Make sure you download the UF that is compatible with the Linux OS kernel version and CPU architecture. At the time of writing, UF version 9.0.1 is compatible with 64-bit Linux distribution kernel versions 3.x, 4.x, and 5.x.

The UF download section of the Splunk download web page (link provided at the beginning of the main section) offers three different software packages (`.rpm`, `.deb`, and `.tgz`) and the installation commands for each package are slightly different. The default installation location is `/opt/splunkforwarder`, which is also referred to as `SPLUNK_HOME`.

Let's look at the commands for installing each UF package.

Command for .rpm

The following command installs the `.rpm` file (**Red Hat Package Manager**) into the default `/opt/splunkforwarder` location:

```
rpm -i splunkforwarder-<...>-linux-2.6-x86_64.rpm
```

Command for .deb

The following command installs the `.deb` (**Debian**) package into the default `/opt/splunkforwarder` location:

```
dpkg -I splunkforwarder-<...>-Linux-<amd64>.deb
```

Command for .tgz (TAR file)

The following command untars the `.tgz` file and installs the UF in the `/opt/splunkforwarder` location:

```
tar xvzf splunkforwarder-<...>-Linux-x86_64.tgz -C /opt
```

After installation, you can go to `/opt/splunkforwarder/bin` and run the `./splunk start -accept-license` command, setting the administrator username and password once you're prompted.

At this point, you will see that the Splunk forwarder has been successfully started and is running. You can issue a `./splunk status` command to check its current status.

At this stage, we have the UF up and running. Now, let's take a look at configuring the UF to forward the data to the indexers. This process is also called **forwarding**.

Configuring forwarding

Forwarding involves configuring the UF to establish connectivity to indexer peers for data transmission. This configuration can be achieved through the `outputs.conf` file, which contains the required indexer peer details. The Splunk CLI has commands to configure forwarding, which, in turn, writes to the same `outputs.conf` file.

Indexers can receive the data on the default TCP 9997 port, which must be enabled by system administrators before the connections from forwarders can be accepted. You need to make sure the network is open between the forwarder and indexers. In real-time Splunk deployments, the organization's network administrator can help establish this. In our scenario, we've set up a standalone Splunk Enterprise instance and a UF to ensure seamless connectivity. This arrangement helps avoid connectivity problems unless port 9997, which needs to be available for listening, is somehow unavailable. Let's take a look at both the Splunk CLI and file configuration approaches, for which you need to supply an administrator account username/password when prompted in the CLI:

1. Log in to the UF host and change the directory to $SPLUNK_HOME/bin (the Linux default is /opt/splunkforwarder/bin).

2. Execute the ./splunk add forward-server <indexer-host:receiving-port> command, as shown here:

    ```
    ./splunk add forward-server 10.1.9.2:9997
    Splunk username: admin
    Password:
    Added forwarding to: 10.1.9.2:9997.
    ```

3. After successfully executing this command, Splunk will write the indexer details to the outputs. conf file inside the $SPLUNK_HOME/etc/system/local/ directory. You need to change directories (cd) and open the outputs.conf file to see its contents:

    ```
    # outputs.conf file contents post command execution
    [tcpout]
    defaultGroup = default-autolb-group

    [tcpout:default-autolb-group]
    server = 10.1.9.2:9997

    [tcpout-server://10.1.9.2:9997]
    ```

4. Now that the indexer details have been added for forwarding, we can verify this by issuing a Splunk CLI command to check the connection status. Change the directory to $SPLUNK_HOME/bin on the forwarder and execute the following command:

    ```
    ./splunk list forward-server
    Active forwards:
            10.1.9.2:9997
    Configured but inactive forwards:
            None
    ```

You will be able to see a similar output if the connection to the indexer has been successfully established. To troubleshoot the issue of missing forwarders, you can investigate the logs of the forwarders. These logs can be found in $SPLUNK_HOME/var/log/splunk/splunkd.log. By searching for ERROR logs within these files, you can identify and address any problems.

In a large environment, often, UFs are centrally managed through a DS for app/add-on deployment. In this case, the UF requires a `deploymentclient.conf` file that contains DS details to establish a connection. We'll look at this in the following section.

Configuring deploymentclient

At this stage, we understand how to configure the DS and install the UF and have enabled data forwarding. To get the latest apps/add-ons from the DS, the UF requires a `deploymentlclient.conf` file, which can be created directly on the filesystem using a text editor. Alternatively, the same can be configured through the Splunk CLI command, which, in turn, writes to the `deploymentclient.conf` file. The CLI approach is the safest option as it avoids typos and other mistakes that might arise by directly editing the file.

Let's look at the Splunk CLI and file approaches for configuring the deployment client:

1. Log in to the UF host and change the directory to `$SPLUNK_HOME/bin` (the Linux default is `/opt/splunkforwarder/bin`).

2. Execute the `./splunk set deploy-poll <DS-host:port>` command to configure `deploymentclient`, as shown here:

    ```
    ./splunk set deploy-poll 10.9.8.7:8089
    ```

3. After successfully executing the command, change the directory to `$SPLUNK_HOME/etc/system/local/` and open `deploymentclient.conf`.

4. You should see the contents of the file, as follows:

    ```
    [target-broker:deploymentServer]
    targetUri = 10.9.8.7:8089
    ```

By default, the forwarder "phones home" DS every 60 seconds. This can be configured by manually editing the `deploymentclient.conf` file followed by restarting the UF.

If there are any changes to DS apps (`$SPLUNK_HOME/etc/deployment-apps/<app_name>`) that have been whitelisted to the forwarder client, they will be downloaded to the forwarder client's `$SPLUNK_HOME/etc/apps/<app_name>` location. In rare cases, such as directly editing `serverclass.conf` file on the DS, the apps might not be deployed to forwarder clients, even though they are phoning home and connectivity to the DS is fine. In that case, issuing the following command on the DS notifies the clients of app changes so that they can be downloaded:

```
$SPLNUK_HOME/bin/splunk reload deploy-server
```

Note that for the `deploymentclient.conf` file's configuration changes to take effect and to establish connectivity between the UF and DS, you should restart the forwarder client. Configuration changes that are made through the Splunk CLI are less likely to need a restart as the forwarder reloads the changes and for any changes made to the config file directly, you might need to restart the forwarder for the changes to take effect.

After the restart, execute the following command on the UF to check the connectivity status:

```
./splunk show deploy-poll
```

To troubleshoot any connectivity or phoning home issues, check out the forwarder's $SPLUNK_HOME/ var/log/splunk/splunkd.log file and search for a phone home pattern.

With that, we have finished configuring forwarding and the deploymentclient.conf file. In the next section, we will look at monitoring forwarders through the **Monitoring Console (MC)** app.

Forwarder monitoring

Monitoring our setup is quite important as the number of forwarders grows, and it can sometimes be very challenging. To aid with this, Splunk offers the out-of-the-box MC app, which contains a dashboard to monitor the forwarders.

By default, the forwarder monitoring feature is disabled in MC and must be enabled by following these steps:

1. Log in to the MC-dedicated Splunk instance and navigate to **Settings | Monitoring Console**.
2. Inside **Monitoring Console**, go to **Settings | Forwarding Monitoring Setup**; you will find that it is disabled. Click **Enable**. By default, **Data Collection Interval** is set to **15 minutes**; leave it as-is and click **Save**, as shown in *Figure 4.5*:

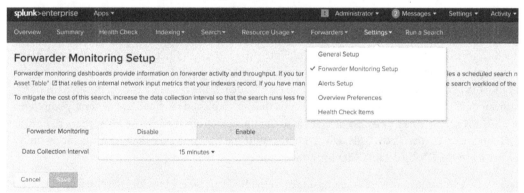

Figure 4.5: Forwarder Monitoring Setup

You will find two menu items under the **Forwarders** tab: **Forwarders Instance** and **Forwarders Deployment**, as shown in *Figure 4.6*. The **Forwarders: Instance** and **Forwarders: Deployment** dashboards are built on the internal logs of forwarders and the DS:

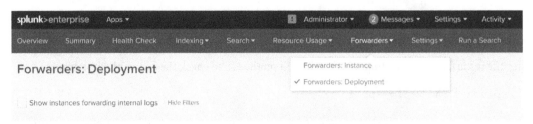

Figure 4.6: Forwarder monitoring dashboards

If the setup for the UF and DS has been done correctly, including the `deploymentclient` configuration, after 15 minutes, you will be able to see that the dashboard panels of **Forwarders: Instance** and **Forwarders: Deployment** have been populated with correct details, such as **Forwarders by Status**, **Forwarders Count**, and **Forwarders Connection count over time**, and so on.

Summary

In conclusion, the Splunk UF is a software binary that does not require any license and is typically installed on the source system. The UF is lightweight and consumes fewer resources. The UF monitors and forwards the data to indexers for indexing and reads the data in a file exactly once using the fishbucket concept. Structured data such as CSV, XML, JSON, and so on can be parsed using `INDEXED_EXTRACTIONS`. Forwarding on the UF is configured in the `outputs.conf` file, which contains the indexer host and management port details.

The UF can be installed on various supported OSs and hardware specifications; we have seen its installation for both Windows (through the interactive GUI) and Linux (through the CLI/silent mode). By default, the UF is installed in `/opt/splunkforwarder/` in linux environment, which is referred to via the `$SPLUNK_HOME` environment variable.

After that, we explored the DS, which is a Splunk Enterprise instance for managing the many forwarders in large Splunk environments. The apps are stored in `$SPLUNK_HOME/etc/deployment-apps` and the location of the DS and the `serverclass.conf` file contains the mapping of apps to deployment clients (that is, forwarders). The apps that are deployed to clients are generally stored in the `$SPLUNK_HOME/etc/apps` directory. To get the apps from the DS, the forwarders require a `deploymentclient.conf` file, which contains the DS host and port details. After successfully connecting to the DS, by default, clients phone home every 60 seconds.

Finally, we went through forwarder management in the MC app, which provides dashboards based on the internal logs of the UF and DS. By default, forwarder management is disabled and can be enabled on the Splunk instance that is dedicated to the MC app in a distributed Splunk architecture. In the next chapter, we will learn about the Splunk configuration files in detail.

In the next section, you will be provided with some questions and answers to test your knowledge of this chapter.

Self-assessment

This section will test you on what you have learned in this chapter. You will be given 10 questions with answers to choose from. The question pattern is the same as what we discussed in *Chapter 1*, in the *Introducing the exam's test pattern* section. After the assessment, refer to the sections where you had difficulty answering questions; alternatively, you could refer to the official Splunk documentation. All the best!

1. The UF can store indexed data. Is this statement true or false?

   A. True.

   B. False.

2. Choose the statements that apply to UFs:

   A. The UF is a lightweight software binary that's typically installed on the source machine.

   B. The UF requires a separate license from Splunk Inc.

   C. The UF can input data and optionally parse structured data (such as CSV, PSV, XML, or JSON).

   D. The UF resends the whole file from the beginning if it gets rebooted.

3. Choose the statements that apply to DSs:

   A. The DS is a Splunk instance and by default, it is enabled.

   B. Apps for forwarders in the DS are stored under `$SPLUNK_HOME/etc/deployment-apps`.

   C. The DS doesn't require connectivity to forwarders. Forwarders reach out to the DS when necessary for config updates.

   D. The DS is an overhead and, without it, many forwarders can be managed easily through external scripting.

4. What is the `.conf` file that the DS maintains for apps and forwarder mapping details?

   A. `deploymentserver.conf`.

   B. `serverclass.conf`.

   C. `deploymentclient.conf`.

   D. `limits.conf`.

5. Which file contains DS details on a forwarder to help establish connectivity with the DS?

    A. `server.conf`.

    B. `web.conf`.

    C. `deploymentclient.conf`.

    D. `serverclass.conf`.

6. Does the forwarder require connectivity to the indexer? If so, which statements are true about it?

    A. No; the forwarder operates independently and does not require connectivity to the indexer.

    B. Yes; the forwarder needs a network connection to the indexer to send data.

    C. Yes; forwarding is enabled on forwarders through the `outputs.conf` file for indexing.

    D. Yes; the forwarder sends data to the indexer on the default `9997` receiving port.

7. Which statement is true about the UF and HF?

    A. The HF cannot be connected to the DS.

    B. The UF is only compatible to work with the DS.

    C. The HF and UF are both able to connect to the DS and get the latest apps.

    D. Forwarders can be restarted through the DS after changes have been made through the `restartSplunkd` setting.

8. What are the ways to monitor forwarders? (Select all that apply)

    A. By writing a script that polls the management port, `8089`.

    B. Through **Forwarder Management** in Splunk Web.

    C. Through the MC app.

    D. Through DS and forwarder internal logs.

9. What is the default "phone home" internal setting on forwarders?

    A. 60 minutes.

    B. 60 seconds.

    C. 300 seconds.

    D. 600 seconds.

10. If you have an app that is specific to Linux forwarders and you need to block it from Windows forwarders, which settings should you choose in `serverclass.conf`?

    A. Allow/deny settings.

    B. Whitelist/blacklist.

    C. Whitelist for Linux hosts and blacklist for Windows hosts on the same server class.

    D. No action – forwarders are intelligent enough to choose the right apps.

I hope you were able to recollect the topics that we went through while going through these questions. We'll review the answers in the next section.

Reviewing answers

Here are the answers to this chapter's questions:

1. *False* – the UF cannot store, search, and index the data. It doesn't have any user interface to configure either.

2. *Options A and C* are the correct answers. By default, the UF comes with an embedded license, so no extra license is required. The UF can parse structured data by setting `INDEXED_EXTRACTIONS` in `props.conf` locally. The UF will resume from the file's last read position by making use of the fishbucket index, which contains seek pointers.

3. *Options A and B* are the correct answers. By default, the DS is enabled on all Splunk instances. Apps are stored in `$SPLUNK_HOME/etc/deployment-apps`. The DS and forwarders always need connectivity and they communicate at defined intervals.

4. *Option B* is correct. `serverclass.conf` is the file that contains the app mapping to forwarder clients and other settings such as `whitelist`, `blacklist`, `restartSplunkd`, and others.

5. *Option C* is correct. `deploymentclient.conf` is the file that contains the DS host and port details to establish a connection.

6. *Options B, C, and D* are the correct answers.

7. *Options C and D* are the correct answers. Both the UF and HF can be managed through the DS. The `restartSplunkd` setting helps reboot the remote forwarder once apps have been deployed.

8. *Options B, C, and D* are the correct answers. The internal logs of the forwarder and DS are used inside the MC app dashboards.

9. *Option B* is correct. 60 seconds is the default phone home interval.

10. *Option C* is correct. Whitelist and blacklist are the two useful settings in such cases.

5

Splunk Index Management

Indexes are repositories of data. Splunk Enterprise stores data as events in indexes. An event refers to a single data record or log entry. It could be a line from a log file, a message from a network source, or any piece of information that is indexed and processed by Splunk. So far in this book, we have seen the forwarders used to monitor and forward data to indexers. You must be wondering how data is processed and where it is stored in the indexer component. In this chapter, you will get the answers you are looking for. It is crucial for system administrators to know about indexes as they organize the creation, management, access control, and storage estimations of indexes in their day-to-day work.

We will begin by learning about Splunk indexes, including default indexes, and how data is organized into buckets with retention policies. After you are familiar with the core concepts, we will move on to bucket types and their rollover behavior, followed by the basic and advanced options pertaining to the creation of indexes. The advanced options relate to the size of bucket directories and which options can be set to improve the performance of large indexes. As the indexes grow over time, it's essential for the administrator to establish a backup strategy. In the event of hardware failures, this guarantees that the indexes can be restored to their original state. Some scenarios require the permanent deletion of data from search activity, which we will go through at a high level. Finally, you will learn how to monitor indexes through the monitoring console app.

In this chapter, we will cover the following topics:

- Understanding Splunk indexes
- Understanding buckets
- Creating Splunk indexes
- Backing up indexes
- Monitoring Splunk indexes

Understanding Splunk indexes

An index is a specific type of data storage inside Splunk Enterprise; in other words, to keep it simple, an index is a repository of data. For example, you are searching for your address on a form online and you input your address details: you provide your house number, your street, and your postcode, which is unique to you. Similarly, if you want to search for specific data in Splunk, you can find it by using its index.

There are two types of Splunk indexes. They are called **event** indexes and **metrics** indexes. Event indexes store any type of text data, and this is the default index type. Metrics indexes only store metrics data, which must comply with a defined structure. There are special commands in Splunk, usually prefixed with m such as `mstats`, `mpreview`, and `mcatalog`, for working with metrics data. Metrics indexes are a completely different topic and beyond the scope of this book. If you would like to read more about them, please visit `https://tinyurl.com/5n8jbd6c`. In this chapter, we will focus on event indexes. From now on, if there is no explicit description of the index type, assume it is an event index.

Splunk allows you to create your own indexes, and it also includes some event indexes, each serving a specific purpose. Let's look at these indexes:

- `main`: The default Splunk index if one isn't specified in external input sources.

- `summary`: A default summary index for storing summary data.

- `_internal`: Stores Splunk internal logs from a range of internal sources, such as splunkd daemon logs, web access logs, scheduler logs, python logs and more.

- `_audit`: Stores Splunk internal audit logs such as logins, searches, data accesses, and administrative activities.

- `_introspection`: Stores Splunk resource consumption and performance data. Monitoring console app dashboards highly depend on this index.

- `_fishbucket`: Stores checkpoint information of file-monitoring inputs.

- `_telemetry`: Stores instrumentation information if enabled through `telemetry.conf`.

There is more than one way to create a Splunk index. It all depends on the type of Splunk deployment architecture. Regardless of the approach, you do the configuration of the `indexes.conf` file in the indexer.

The indexer is a Splunk Enterprise instance that processes incoming data before storing it in indexes and responds to search queries issued by a search head. We had a brief overview of Splunk components such as the indexer and search heads in previous chapters.

Let's go through the important characteristics of indexes:

- Splunk allows administrators to create custom indexes to store and organize specific types of data based on their needs and requirements.

- Every index has data size and retention time limitations. To view the configuration settings, see the `indexes.conf` definition of `_internal` index in the `$SPLNUNK_HOME/etc/system/default` directory.

- Indexes contain originally received data in a compressed format, which is called raw data. They also contain index files called `tsidx` files and `metadata` files.

- Indexes organize data in the form of buckets on the filesystem.

- Indexes are by default stored at `$SPLUNK_HOME/var/lib/splunk`. The path is referred to with the `$SPLUNK_DB` environment variable.

- Once the data is written to an index, it is final and cannot be modified.

- An authorized user that has the `can_delete` role can issue a `delete` command to prevent data from participating in a search. This command applies delete markers to data. This means it doesn't permanently remove the data from storage. To permanently delete event data, execute the `splunk clean` command with caution.

- The segregation of indexes depends on data size, data retention, access policies, and the type of data being stored in them.

I hope you now have enough of an understanding of indexes. Let's go through how Splunk organizes data in the form of buckets in the next section.

Understanding buckets

Buckets are an integral part of indexes; they contain raw data and index files. They are organized in the form of folders on a filesystem with a specific naming pattern. These folders are explicitly used by the Splunk indexer for data storage and search processing.

In order to learn a bit more about them, let's look at the default `_introspection` index folder structure:

```
C:\Program Files\Splunk\var\lib\splunk\_introspection>tree
Folder PATH listing for volume Windows
Volume serial number is 6676-C13F
C:.
├───colddb
├───datamodel_summary
├───db
│   ├───db_1663851317_1661686760_13
│   │   └───rawdata
│   ├───GlobalMetaData
│   ├───hot_v1_14
│   │   └───rawdata
└───thaweddb
```

Figure 5.1: Splunk non-clustered index folder structure

Figure 5.1 shows the `_introspection` index inside the `$SPLUNK_DB` path. The naming convention is only applicable to non-clustered indexers. Let's take a look at the `indexes.conf` file located in the `$SPLUNK_HOME/etc/system/default/` directory. It contains the `_introspection` index settings that correlate with the folder structure in *Figure 5.1*:

```
# indexes.conf - _introspection internal index settings
[_introspection]
homePath = $SPLUNK_DB\_introspection\db
coldPath = $SPLUNK_DB\_introspection\colddb
thawedPath = $SPLUNK_DB\_introspection\thaweddb
coldToFrozenDir = $SPLUNK_DB\_introspection\frozenDir
maxDataSize = 1024
frozenTimePeriodInSecs = 1209600
```

In this configuration of the `_introspection` index, while the term "buckets" is not explicitly used, we can establish an equivalent mapping to the following settings:

- `homePath` stores hot and warm buckets

- `coldPath` stores cold buckets

- `thawedPath` stores preprocessed data for re-indexing

Let's go through buckets and index settings in detail.

Splunk stores data in more than one type of bucket. They are all given the suffix `db` inside the configuration. The types of buckets are `Hot`, `Warm`, `Cold`, `Frozen`, and `Thawed`. The transition of buckets is shown in the following diagram:

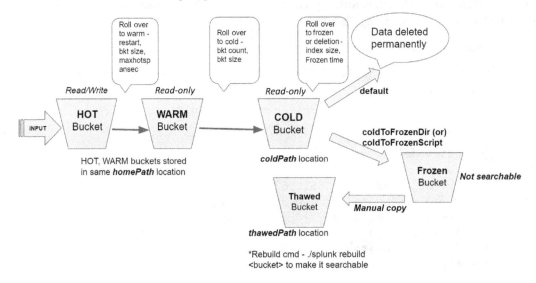

Figure 5.2: Splunk index buckets transition

Figure 5.2 shows the buckets inside an index, how the data is organized, and the purpose of each of them.

Let's go through the bucket types and how they work:

- **Hot**: Hot buckets contain freshly arrived data. Before it is stored, data is parsed in the parsing phase and goes through the license meter. Hot buckets are writable and readable simultaneously. In the index definition, hot buckets are specified to store in the `homePath` location. The folder has the prefix `hot*`; as you can see in *Figure 5.1*, the example hot bucket is `hot_v1_14`.

- **Warm**: Hot buckets are rolled over into warm in three scenarios: when the indexer is rebooted, when the hot bucket reaches the maximum size (`maxDataSize`), or when the maximum lifespan of the bucket is reached (`maxHotSpanSecs`). Warm buckets are stored in the same location as hot buckets, `homePath`, and they are only readable. The folder is named using this convention: `db_<earliesttime>_<latesttime>_uniqueid`. As you saw in *Figure 5.1*, `db_1663851317_1661686760_13` is a warm bucket. Here, `<earliesttime>` (`1663851317`) and `<latesttime>` (`1661686760`) are in epoch time format, and `13` is the unique ID. When performing a search in Splunk, the search process determines which buckets to open and search for event matches based on the specified search time range (`earliesttime`, `latesttime`).

- **Cold**: Splunk moves warm buckets into `coldPath` when they reach certain criteria, and bucket names remain the same after they move. The oldest warm buckets are moved first when the `homePath` location reaches its defined `maxDataSizeMB` or if the number of warm buckets reaches the defined `maxWarmDBCount` setting. The data in cold buckets is readable and searchable.

- **Frozen**: Frozen buckets are completely optional and are configured when the administrators choose to retain data from cold buckets. Without frozen bucket configuration, data from cold buckets will be deleted after it reaches a certain age. Frozen buckets are not searchable. You can specify either `coldToFrozenScript` or `coldToFrozenDir` to move raw data from a cold bucket to a frozen bucket. If neither of these options is specified, then the data in the cold bucket will be permanently deleted. If you specify both options, `coldToFrozenDir` takes precedence and `coldToFrozenScript` is ignored. The criteria for deleting or moving data from a cold bucket to a frozen bucket is either when the data in the cold bucket reaches the age specified in the `frozenTimePeriodInSecs` option or when the index's maximum total data size is reached. The age of the bucket takes precedence over the size of the index when moving or deleting data from a cold bucket.

- **Thawed**: This is also an optional bucket and it works in coordination with the `Frozen` bucket. We have seen that frozen buckets are not searchable. To make data in frozen buckets searchable, data from `frozen buckets` must be copied to `thawedPath` and rebuilt using the `./splunk rebuild <bucket>` command. Data can be retained in `thawedPath` as long as it needs to be searched, and no retention policies apply. This means it will never be deleted by Splunk Enterprise. If the data is no longer required for searching, then the administrator can remove it from `thawedPath`.

We've now looked at event index buckets. Data gets rolled from hot to warm to cold to frozen to thawed buckets depending on the situation. Now that you are familiar with indexes and their internal structure, let's move on to the creation of indexes.

Creating Splunk indexes

Creating a custom index is a Splunk administrator's responsibility. As you have seen, data is retained in indexes until it reaches a certain age or the index reaches a certain size, so an estimation of an index's size before it is created is a crucial step. Having an accurate estimation retains the data for the necessary amount of time, and it also helps to determine the required storage capacity upfront.

In order to estimate the size of the index, we must have the answers to the following questions:

- *Question*: How long does the data need to be retained in days?

  Answer: For example, 100 days

- *Question*: How much data volume per day is expected in GB?

  Answer: For example, 1 GB

The formula for index size estimation is as follows:

Retain days X volume per day X 1/2 (raw data compression + index files)

If we substitute in the example values, the index size becomes *100 X 1 X (½) = 50 GB*.

Now that you know how to estimate the size of an index and have decided on a name for it, let's go through the index creation process.

In the *Understanding buckets* section, we saw that the indexes.conf file in the $SPLUNK_HOME/etc/system/default directory has an entry for the _introspection index. Similarly, Splunk ships with many internal indexes that you can find in the same file. Splunk allows us to create custom indexes, and there is more than one way to create an index. It all depends on the type of deployment architecture. In general, the administrator chooses the most suitable approach. In this section, we have chosen a single instance of Splunk, which serves as both the search head and the indexer. Let's take a look at the different ways to create event indexes.

Splunk Web

Log in to Splunk Web as an administrator. Go to the **Settings** menu in the top right and click on **Indexes**. A page opens with a **New Index** button in the top right. On the same page, you can see the default indexes are prefixed with _, such as _internal and _audit.

If you wish to see the configuration for one of the default indexes, click on the index name. Otherwise, proceed with creating a new index by clicking the **New Index** button and filling in the **Index Name** and **Max Size of Entire Index** fields. The default **Index Data Type** setting is **Events**. Metrics should be explicitly selected if you want to create a metrics index. I have set the index name to `windows-event-logs` and set the size to `500` MB. Then, click the **Save** button.

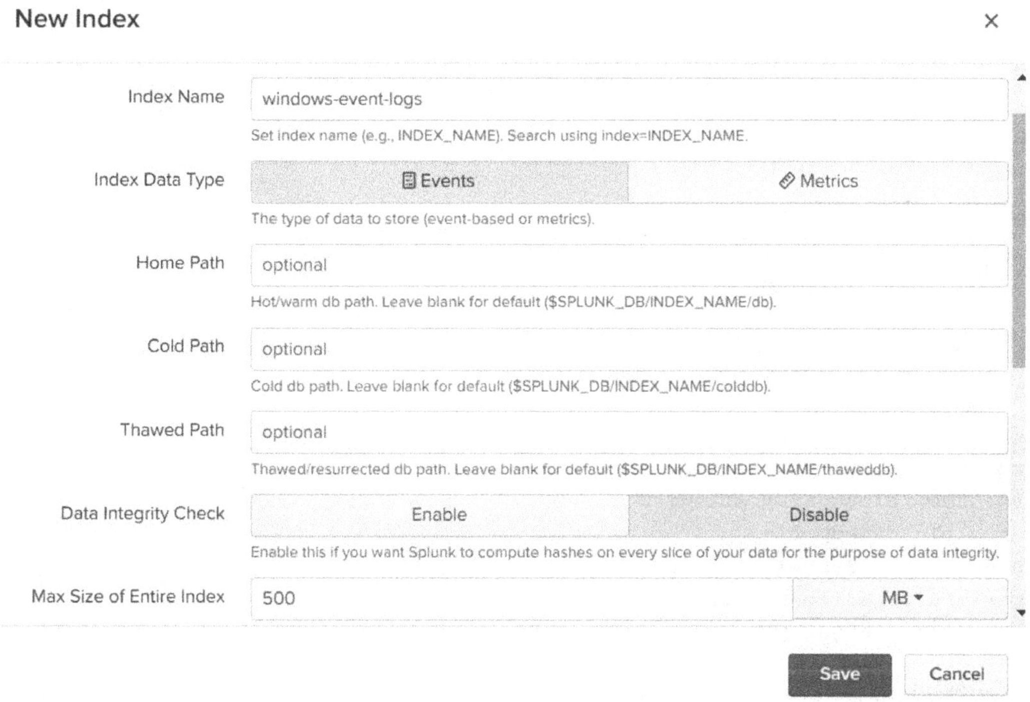

Figure 5.3: New index creation form

While creating the new index in *Figure 5.3*, Splunk automatically chooses the default `$SPLUNK_DB` location and the `Search & Reporting` app, along with other default settings such as **Max Size of Hot/Warm/Cold Bucket** and **Data Integrity Check** During testing, it is often acceptable to retain these default settings and proceed to the next step. Now, let's explore the **Command-Line Interface (CLI)** approach.

CLI

The Splunk CLI offers a variety of commands, and adding indexes is one of them. You can log in to your Splunk instance through SSH if the OS is Linux based. For Windows, you can use Command Prompt. Let's take a quick look at how to add indexes using Splunk CLI commands in the Windows environment:

- **Syntax:** `$SPLUNK_HOME\bin\splunk add index <index_name> -<parameter> <value>`

- **Parameters**: `homePath`, `coldPath`, `thawedPath`, `app`, and so on

- **Example**: `$SPLUNK_HOME\bin\splunk add index windows-event-logs -app search`

Both the CLI and Splunk Web approaches write to the `indexes.conf` file on the filesystem. If you are familiar with the `indexes.conf` specification, then creating an `indexes.conf` file directly achieves the same purpose.

indexes.conf explained

Let's deep dive into the `indexes.conf` file for the `windows-event-logs` index that we have created through Splunk Web. As we learned in the *Splunk Web* section, the index has by default been saved to the Search & Reporting app folder called `search`, hence I can find it at `$SPLUNK_HOME/etc/apps/search/local`:

```
[windows-event-logs]
coldPath = $SPLUNK_DB\windows-event-logs\colddb
enableDataIntegrityControl = 0
enableTsidxReduction = 0
homePath = $SPLUNK_DB\windows-event-logs\db
maxTotalDataSizeMB = 500
thawedPath = $SPLUNK_DB\windows-event-logs\thaweddb
```

One of the advantages of this approach is there is a range of options that can be set for advanced index configurations. However, the creation of a new index through this file-based approach requires the Splunk instance to be restarted.

There are ways to control the size of data from the bucket directory level to the total index size. Let's take a look at the following advanced options and what to back up:

- `homePath.maxDataSizeMB` limits the `homePath` directory size for hot and warm buckets.

- `coldPath.maxDataSizeMB` limits the `coldPath` directory size for cold buckets.

- To limit the number of hot and warm buckets that can be created, set `maxHotBuckets` and `maxWarmDBCount`, respectively. For high-volume indexes, where the indexing size is more than 10 GB/day, Splunk suggests increasing `maxHotBuckets` from the default of 3 to 10.

- By default, Splunk limits the hot bucket size to 750 MB through `maxDataSize = auto`, and if the index size per day estimation is 10 GB, set this option to `maxDataSize = auto_high_volume` for Splunk to create 10 GB hot buckets on 64-bit systems. The reason for having this option is to avoid creating too many small buckets for large indexes. Having too many small buckets affects Splunk's performance. Hot buckets will roll to warm buckets once they reach their maximum size and then a new hot bucket will be created.

- `maxTotalDataSizeMB` is the total index size limit. If you look at the example index, `windows-event-logs`, we have set this value to 500 MB.

- Global settings are applied to every index specified in `indexes.conf`. For a complete list, refer to the `indexes.conf.spec` file in the Splunk docs. At the global level, the `lastChanceIndex` option can be configured. This acts as a final option for any external inputs that don't indicate a valid index. If this value is left empty, Splunk will simply discard these events.

What if the data moved to cold buckets is not being searched very frequently, so the allocation of expensive hardware storage similar to hot/warm buckets isn't cost-effective? Splunk suggests using low-latency high-throughput storage for hot/warm buckets as they hold recent data that is searched quite often. Cold buckets can reside on standard latency storage, which is slow and cheap. To achieve this, an index can be configured using volumes. Let's look at volumes in the next section.

Index volumes

Volumes are specific directories on a filesystem that are allocated for indexes in order to manage storage space effectively. Let's take a look at the volume configuration in a UNIX-based OS for the `windows-event-logs` index that we created on Splunk Web in the previous section:

```
[volume:hot_storage]
path = /mnt/fast_disk
# Optional limits the volume size to 60 GB
maxVolumeDataSizeMB = 61440

[volume:cold_storage]
path = /mnt/slow_disk
#Optional limits the volume size to 50 GB
maxVolumeDataSizeMB = 500

#index definition
[windows-event-logs]
homePath = volume:hot_storage/windows-event-logs
coldPath = volume:cold_storage/windows-event-logs
maxTotalDataSizeMB = 51200
thawedPath = $SPLUNK_DB\windows-event-logs\thaweddb
```

This configuration contains the `volume` keyword in `homePath` and `coldPath`. `thawedPath` is set to the standard location as it cannot be on a volume. If you would like to know some more advanced options, then look at the `indexes.conf.spec` file in the `$SPLUNK_HOME/etc/system/README` directory.

In the next section, we will go through backing up indexes.

Backing up indexes

So far, we have created indexes and stored information in buckets, and modified their configurations in the .conf file. What if the underlying hardware fails and you have critical data and files that cannot be lost? To restore your Splunk instance to its original state, a backup procedure must be set up. For the certification exam, it is important to know which folders of the Splunk installation to back up. The following two essential folders need to be backed up:

- The $SPLUNK_DB directory: $SPLUNK_HOME/var/lib/splunk/

  Hot buckets cannot be backed up while Splunk is running; instead, a snapshot can be taken incrementally, and you can take a backup of the snapshot

- The $SPLUNK_HOME/etc directory: This contains apps, user configurations, system configuration files, and licenses

In Splunk, the $SPLUNK_HOME/etc directory is a critical directory that contains configuration files and settings that control the behavior of the entire Splunk deployment. Backing up this directory is essential to ensure that all configurations and customizations are safeguarded and can be restored in case of any unexpected issues or system failures.

That is everything we're going to cover about the creation of indexes; we ended with what to back up. At this stage, you should be familiar with core index concepts and advanced configuration options. Let's take a look at monitoring indexes in the following section.

Monitoring Splunk indexes

As an administrator, it can be overwhelming to monitor several indexes as they grow in number. To address this, Splunk monitors indexes through a default monitoring console app. It has many monitoring features, and monitoring indexes is one of them. Let's take a look at its menu options.

Log in as an administrator and navigate to **Settings** in the top right. Click on **Monitoring Console**. The console opens in the following web view. Click on **Indexing** and select **Index Detail: Instance**.

Figure 5.4: Indexes dashboard in the monitoring console

As you can see in *Figure 5.4*, Splunk is a standalone Windows-based instance that has three dashboards for monitoring indexes and their volumes. The **Index Detail: Instance** dashboard has the following settings:

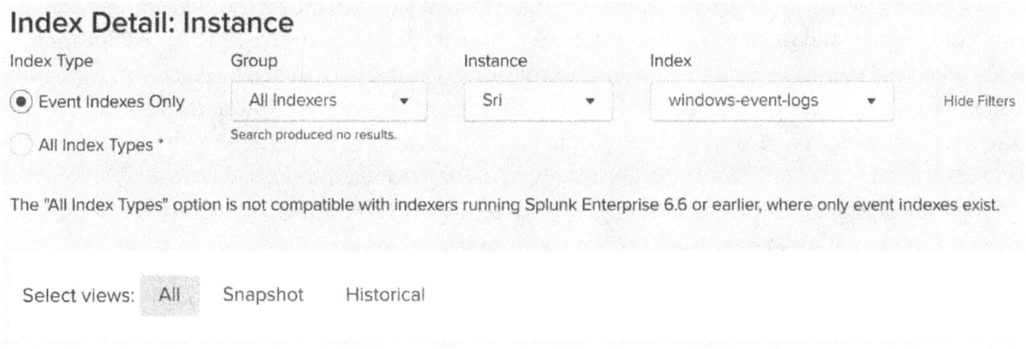

Index Detail: Instance

Index Type	Group	Instance	Index	
● Event Indexes Only	All Indexers ▾	Sri ▾	windows-event-logs ▾	Hide Filters
○ All Index Types *	Search produced no results.			

The "All Index Types" option is not compatible with indexers running Splunk Enterprise 6.6 or earlier, where only event indexes exist.

Select views: All   Snapshot   Historical

Figure 5.5: Index Detail: Instance dashboard

The dashboard has drop-down options for **Instance** (also called the Splunk indexer), and the index that has been chosen is `windows-event-logs`. Upon selection of all the options, the dashboard panels will refresh and show the status of `homePath`, `coldPath`, index size by usage, volume details, and bucket details, including the count and number of events per bucket. To look at all the indexes in one dashboard view, choose the **Indexes and Volumes: Instance** menu option. Similarly, you can explore other dashboards and get an idea of the key performance and usage metrics of indexes, buckets, and volumes.

The monitoring console is a one-stop application in Splunk for monitoring indexes. With this, we have come to the end of the chapter.

In the next section, you will be given questions and answers that are similar to the exam questions for you to test yourself on index management.

Summary

Index is a name given to a specific data repository in Splunk. An index can be configured with simple basic settings and some advanced settings as it grows larger. We have learned that there are two index types: event indexes and metrics indexes. Event indexes can store any text data, whereas metrics indexes store data that follows a specific metric structure. You learned about destructive commands, **delete** and **clean**. These commands should always be used with extreme caution and should only be executed when you are absolutely certain of their implications. To delete an index, we can use a special `can_delete` role, which applies delete markers to data without removing data from storage. Through the CLI, data can be deleted permanently using Splunk's `clean` command.

We also explored the role of Splunk indexers as crucial components of the Splunk architecture. Indexers are responsible for processing and indexing data, storing it efficiently, and responding to search requests from users and external applications. Data is organized into buckets in indexes, which are called hot, warm, cold, frozen, and thawed. These buckets are stored by default in the $SPLUNK_DB location. We discussed the transition from hot to warm to cold and, optionally, to frozen buckets when criteria are met. Frozen buckets are not searchable; thawed buckets are used to restore data from frozen buckets for searching. We also went through the creation of indexes through the web, the CLI, and the indexes.conf file. We looked at the advanced settings for data aging, performance, volumes for storage management, and so on. One of the global settings is lastChanceIndex, which is used to store external inputs that have been supplied with an invalid index.

We discussed backing up index data and other useful configurations for restoration. We ended by discussing the monitoring console app, which contains dashboards for monitoring indexes and volumes. In the next chapter, we will dive into the Splunk configuration files.

Self-assessment

You will be given 10 multiple-choice questions. The question patterns are the same as discussed in the *Introducing the exam's test pattern* section of *Chapter 1*, *Getting Started with the Splunk Enterprise Certified Admin Exam*. When you have finished, refer to the sections with which you are having difficulty. Alternatively, you can refer to the Splunk documentation. All the best! Let's get started.

1. An index is a repository of data, and it has a name. Is this statement true or false?.

    A. True

    B. False

2. An indexer is a Splunk instance. What is its role in a Splunk deployment? (Choose all that apply)

    A. The indexer stores data.

    B. The indexer stores data in $SPLUNK_HOME location.

    C. The indexer responds to search requests issued by the search head.

    D. Data stored in the indexer cannot be deleted.

    E. The indexer by default comes with internal indexes. You can create custom indexes too.

3. Do metrics indexes store any type of text data?

    A. Yes

    B. No

4. Choose the Splunk indexes that are included by default in the Splunk installation.

    A. `_internet`

    B. `_audit`

    C. `_telemetry`

    D. `summary`

    E. `main`

5. Which of the following statements is true about the `maxDataSize` setting in `indexes.conf`?

    A. Setting `maxDataSize = 750MB` sets the index size to 750 MB

    B. Setting `maxDataSize = auto` sets the hot bucket size to 750 MB

    C. Setting `maxDataSize = 10GB` sets the index size to 10 GB

    D. None of the above

6. Can you identify the names of the buckets that Splunk index data is organized into?

    A. Hot, cold, warm, freeze, through

    B. Hot, warm, cool, freeze, thaw

    C. Host, warm, cold, frozen, thawed

    D. Hot, warm, cold, frozen, thawed

7. Which statements are true about hot and warm buckets?

    A. Hot buckets are first in line to receive the data copy for storage.

    B. Hot buckets roll over into warm buckets.

    C. Hot buckets are readable and writable.

    D. Warm buckets are only readable.

8. You are a system administrator and your manager advised you to clean an index that you don't want to permanently delete. What actions would you perform?

    A. Log in to Splunk Web and execute a `delete` operation on the **Settings** page.

    B. Log in to the Splunk Web home page of the indexer and execute a `clean` command.

    C. Log in to the Splunk Web instance that has a distributed search connection to the indexer and execute a search with `index=<index-name> | delete` command.

    D. Log in to the Splunk instance through the CLI and execute `./splunk delete <index-name>`.

9. Which statement is true about volumes in `indexes.conf`? (Choose all that apply)

    A. Volumes are directories on a filesystem.

    B. Volumes are complex to manage and hence require Splunk support assistance.

    C. Volumes are useful for managing storage effectively.

    D. Volumes are configured using stanzas with the prefix `volume:`.

10. Can you locate the directory to be backed up regularly?

    A. `$SPLUNK_HOME/var/etc`

    B. `$SPLUNK_DB/etc`

    C. `$SPLUNK_HOME/etc`

    D. `$SPLUNK_DB/var/lib/splunk/fishbucket`

I hope you were able to recollect the topics! Let's look at the answers in the next section.

Reviewing answers

1. *True*, indexes are repositories of data and they do have names.

2. *Option A, C,* and *E* are the correct answers. Indexer data is by default stored in the `$SPLUNK_DB` location. Indexed data can be deleted through the `delete` and `clean` commands.

3. *No*, metrics indexes store only metrics data in text form.

4. *Option B, C, D,* and *E* are the correct answers.

5. *Option B* is correct. `maxDataSize = auto` sets the hot bucket size to 750 MB. The other option you can set is `maxDataSize = <integer> | auto | auto_high_volume`. Just specify the integer (`maxDataSize = 1000`) or `auto_high_volume` sets the bucket size to 10 GB, which is preferred for high-volume indexes with a daily data volume over 10 GB.

6. *Option D* is the correct answer. The data rolls over from hot to warm to cold to frozen (optional). Thawed buckets are for restoring frozen data.

7. All options are correct.

8. *Option C* is correct. For example, `index=web-logs | delete` applies delete markers to the data that has matched search conditions and it doesn't delete the index data permanently from the disk.

9. *Option A, C,* and *D* are the correct answers.

10. *Option C* is the correct answer. The `$SPLUNK_HOME/etc` directory contains critical information about users, apps, and other system configuration details.

Splunk Configuration Files

Configuration files are an integral part of Splunk Enterprise. Any configuration change that you do via Splunk Web, the CLI, or a RESTful interface will get updated to a conf file directly behind the scenes. Henceforth, whenever encountering terms such as ".conf," "conf," "config," or "configuration file," they should all be understood to refer to the same concept. There are many such conf files that exist to fulfill a specific purpose. In this chapter, you will find examples of the usage of commonly used files across Splunk components, such as the search head, indexer, and forwarder. These files can be stored in multiple places, such as in user, system, and app directories. You will learn how these files get merged in memory for processing. As you get familiar with the directory structure where the .conf files can be stored, you'll come to understand their precedence, the order in which they get processed, and find them using the btool command.

You will learn that .conf files are highly important to both Splunk system administrators and data administrators. System admins often need to tune or re-configure system-level settings, and data admins do ingestion, search-time, index-time knowledge object creation, and troubleshooting of data indexing. Knowing how the conf files work together and seeing the merged conf files through btool and a RESTful interface is necessary for everyone to receive more insights into Splunk.

In this chapter, we will cover the following topics:

- Understanding conf files
- Understanding conf file precedence
- Troubleshooting conf files

Understanding conf files

.conf files in Splunk Enterprise are configuration files that serve specific functions during the runtime of data forwarding, parsing, indexing, and searching. These files contain important settings and can be modified through the filesystem, either directly or indirectly via Splunk Web, the CLI, or a REST API.

Let's explore key information about .conf files.

File format and access

To modify or create new `.conf` files, you need access to the Splunk component filesystem. This typically requires appropriate permissions or administrative privileges on the machine where Splunk is installed. By accessing the filesystem, you can edit the `.conf` files using a text editor or other tools.

Structure and syntax

The `.conf` files use the **<key/attribute>** = **<value>** format to define settings. Each key has its own significance, and keys are case-sensitive. Settings are organized into unique [`<stanza>`] sections. It's important to follow the defined specification for each `.conf` file, which can be found in Splunk's online documentation or the installation directory under `$SPLUNK_HOME/etc/system/README` with a `.spec` extension.

For example, in the `props.conf` file, KV_MODE is the correct attribute, and kv_mode is treated as a different and incorrect attribute. Make sure to use the correct casing when defining attributes in `.conf` files by following the specification of the respective conf file. The following is an example structure of a `props.conf` file containing the [`default`] stanza and other customized stanzas:

```
# This is a comment in .conf files
# Global settings applied to all stanzas in props.conf
[default]
TIME_FORMAT = %Y-%m-%d %H:%M:%S
TRUNCATE = 20000
CHARSET = UTF-8


# Specific settings for a sourcetype
[access_logs]
EXTRACT-field_name = ^\[(\d{4}-\d{2}-\d{2} \d{2}:\d{2}:\d{2})\]


# Settings for a host
[host::server1]
TRANSFORMS-sethost = sethost_server1
LINE_BREAKER = ([\r\n]+)(\d{4}-\d{2}-\d{2})
EXTRACT-myfield1 = myregex1
EXTRACT-myfield2 = myregex2
```

Here are some notes on this example:

- Comments can be added using the # symbol at the beginning of a line

- The [`default`] stanza sets global settings that apply to all source types unless overridden

- The [`access_logs`] stanza defines specific settings for the access_logs source type

- The [`host::server1`] stanza specifies settings for a particular host named server1

- The `LINE_BREAKER` setting defines a regular expression pattern to break events into individual lines

- The `EXTRACT` settings define field extractions using regular expressions

- Each setting is defined as a `<key>` = `<value>` pair

This is just a basic example to illustrate the structure of a `props.conf` file. The actual configuration settings and stanzas may vary, depending on your specific requirements and the data sources you work with.

Config layering and inheritance

Splunk supports config layering and inheritance, allowing settings to be inherited from higher-level stanzas by lower-level stanzas. This simplifies configuration management by reducing duplication. The exact inheritance rules depend on the specific `.conf` file and its purpose. Let's explore an example of conf layering and inheritance using the `inputs.conf` file.

Let's suppose the default `inputs.conf` file contains the following global setting:

```
# inputs.conf default stanza
[default]
host = default-hostname
sourcetype = default-sourcetype
```

Now, in the local `inputs.conf` file, you want to override the host setting for a specific input source. You can add the following overriding stanza:

```
# Unix-like systems - inputs.conf file monitor stanza
[monitor:///path/to/file.log]
host = specific-hostname
```

In this example, the global setting in the default stanza sets the host to `default-hostname` and the source type to `default-sourcetype` for all data inputs. However, the overriding stanza in the local `inputs.conf` file specifies a different host, `specific-hostname`, specifically for the `/path/to/file.log` input source. This overriding setting takes precedence over the global setting for that specific input source. By using this approach, you can define common settings in the default stanza and then override those settings for specific inputs in more specific stanzas. This allows you to inherit the default settings for most inputs while customizing certain inputs as needed.

Default stanzas and global settings

`.conf` files often include default stanzas that define global settings. These settings apply to all instances of a particular component unless overridden by more specific stanzas. Default stanzas are typically found in the `$SPLUNK_HOME/etc/system/default` directory and should not be modified or removed. To customize settings, create new files under either `$SPLUNK_HOME/etc/system/local` or `$SPLUNK_HOME/etc/apps/<app_name>/local`.

Merging multiple conf files

In Splunk, `.conf` files are often found in multiple locations within the parent directory, `$SPLUNK_HOME/etc`, under users, apps, and system subdirectories. When Splunk starts or reloads configurations, it merges multiple `.conf` files of the same type. This merging process combines the settings from different `.conf` files into a single in-memory representation for execution.

Splunk evaluates a specific order for the merging of the same conf file types, dependent on the context of the conf file, which could be **Global** (index time) or the **App/User** (search time). Understanding the order of precedence of `.conf` files is crucial to manage and customize the configuration settings in Splunk effectively. In the upcoming section, we will explore and learn about the precedence order of `.conf` files at search time and index time in more detail.

Let's take a look at the popular configuration files in different phases of Splunk data processing:

- `inputs.conf`: The `inputs.conf` file configures data inputs for monitoring and indexing in Splunk. It is a standard file present in every Splunk instance, primarily used to monitor internal logs such as `splunkd.log`, `metrics.log`, `splunkd_access.log`, and `audit.log` in the `$SPLUNK_HOME/var/log/splunk` directory.

- `props.conf`: This file contains settings applied during the parsing phase of data processing, before indexing and search time field extraction processing.

- `transforms.conf`: This file contains settings to transform data before indexing, advanced field extractions, lookup file settings, and search time field extraction processing

- `indexes.conf`: This file contains index configuration settings.

- `server.conf`: This file contains settings pertaining to server configurations such as SSL, HTTP, and clustering mode.

- `outputs.conf`: This file contains settings related to where to forward data. This is usually present in forwarders and on Splunk components where forwarding is enabled.

To find the full list of configuration files, go to the following Splunk documentation link: `https://docs.splunk.com/Documentation/Splunk/9.0.2/Admin/Listofconfigurationfiles`.

Let's look at the location of `.conf` files that are, in general, placed in any Splunk instance. In the following figure, the `SPLUNK_HOME` variable refers to the `C:\PROGRAM FILES\Splunk` location on a Windows system where Splunk is installed and running:

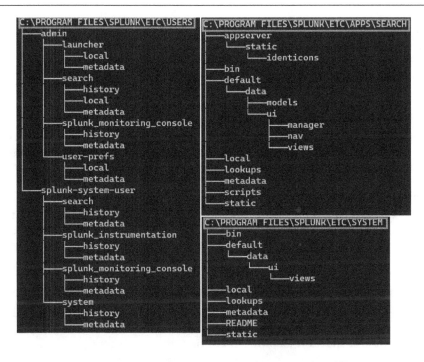

Figure 6.1: The directory structure and location of conf files

Figure 6.1 shows conf files in the SPLUNK_HOME etc\users\<user_name>\<app_name>, etc\apps\<app_name>\, and etc\system directories. The subdirectories inside of them, \default and \local, are the typical locations for .conf files:

- <user_name> in the figure shows the built-in admin and splunk-system-user users

- <app_name> in the figure shows the out-of-the-box search app and its subdirectory structure

Splunk suggests not altering the conf files in the etc\system\default directory. Instead, create an etc\system\local\ directory and place customized configs in there.

Points to remember

Splunk defines multiple configuration files, which are accompanied by their corresponding specifications with the .spec extension. These resources can be found within the $SPLUNK_HOME/etc/system/README directory, along with illustrative examples.

Each and every conf file has a basic structure, having [<stanza>] and settings underneath as <key> = <value> pairs. Keys are case-sensitive.

Splunk ships with default conf files under the $SPLUNK_HOME/etc/system/default directory, and it is not recommended to edit the files inside of this directory. Hence, a /local directory is preferred to override the settings.

The precedence of `.conf` files in different directories determines which settings take priority. Let's explore the order in which these configuration files are considered.

Understanding conf file precedence

`$SPLUNK_HOME/etc` is the parent directory where all conf files exist in a typical Splunk Enterprise installation. A configuration file can be created in more than one place under this parent directory. You might be wondering why there is a need to have the same type of file in multiple places.

The ability to configure settings at various levels within Splunk provides flexibility to administrators, developers, and users to customize the platform according to their specific needs. This flexibility allows for precise control over individual apps and user experiences. The precedence of files that have the same stanza names in multiple directories is determined by Splunk. Precedence is covered later in this section.

By default, Splunk Enterprise ships all the system-wide configurations under the `$SPLUNK_HOME/etc/system/default` directory, which is not supposed to be altered. So, it is suggested that if a change to a conf file under the `/default` directory is required, create it under the `$SPLUNK_HOME/etc/system/local` directory, which takes precedence over the `/default` directory. Similarly, conf files could be present under the following directories:

- `$SPLUNK_HOME/etc/users`: This directory contains private, user-specific changes that are owned by each individual user. This ownership ensures that user-specific changes are isolated and not shared with other users by default.

- `$SPLUNK_HOME/etc/apps/<app_name>/[default|local]`: Configurations created within an app's `<app_name>` directory are stored here. These configurations are available to any user who has access to the app. By assigning specific roles to users, you can control their access to apps, including the ability to view and modify the configurations within the `<app_name>` directory.

- `$SPLUNK_HOME/etc/system/local`: The configurations stored in this Splunk directory are considered global configurations. These configurations are accessible to every user in the Splunk instance.

At this stage, you are aware of the list of configuration file directories. Let's understand how Splunk decides the precedence of the same conf files in different directories (refer to *Figure 6.1*). The precedence is decided based on the conf type input, props, indexes, and so on, and their context.

The context in which a configuration setting is used can also affect the order of precedence. For example, some configuration settings may only apply to a specific app or data source, while others may apply globally to the entire Splunk environment. The two types of context are as follows:

- **Global**: Configuration files that are independent of user and app contexts include settings related to indexing, monitoring, deployment, server configurations, and so on. These configurations are independent of user and app context and apply globally to all users in the Splunk instance.

- **App/user**: Search-time configuration files are stored within the context of an app/user. These files encompass configurations related to dashboards, reports, alerts, field extractions, and other associated settings that are specific to the app/user context.

A list of conf files that are considered under the global and app/user context is available in the Splunk docs: `https://tinyurl.com/24w6z227`.

In the next section, we will explore precedence.

Search-time precedence

Search-time configurations that are scoped under the app/user context are treated according to the order of precedence. Examples of such conf files include `savedsearches.conf`, `macros.conf`, `props.conf`, and others that are relevant during search time. To give an example of scope, a user can create an alert and scope it to run in private mode. This means that the alert will only be accessible and will operate within the resource constraints assigned to the user.

Similarly, an admin has the capability to create a report specifically for project users and scope it to their dedicated app. This ensures that the report is accessible and available only within the context of that particular project's app. Other users outside the project will not have access to this report.

During search time, Splunk evaluates configurations in the following order:

1. `$SPLUNK_HOME/etc/users/<user_name>/<app_name>/local`
2. `$SPLUNK_HOME/etc/apps/<app_name>/local`
3. `$SPLUNK_HOME/etc/apps/<app_name>/default`
4. `$SPLUNK_HOME/etc/system/local`
5. `$SPLUNK_HOME/etc/system/default`

We will look at a couple of scenarios in the following two subsections to understand how the scope of a macro knowledge object defined under the app/user context takes precedence over others. Before we dive into scenarios, let's understand how macros work in Splunk.

In Splunk, macros are reusable snippets of search commands or search strings that you can define and save to make your searches more efficient and easier to manage. You can define macros in both the user context and the app context.

The following two scenarios are about analyzing website performance metrics. Imagine you're part of a web operations team responsible for monitoring and optimizing the performance of your company's website. You use Splunk to analyze website access logs and identify areas for improvement.

Scenario 1

Let's consider an example where you have macros with the same name but defined in different locations within Splunk, one in the `user` directory and the other in the `app` directory. Having macros in different places – user and app directories – strikes a balance. It lets users customize while benefiting from app-provided macros. This way, user changes don't disrupt apps or global setups, making Splunk organized and smooth.

In your personal user context, you might create the following macro named `high_response_time` to quickly filter for events where the response time exceeds a certain threshold. This user macro helps you easily find access logs with response times above 500 milliseconds, allowing you to investigate potential performance issues during your individual analysis.

In the Splunk user directory, the configuration is usually stored in the configuration file called `macros.conf`, located in `$SPLUNK_HOME/etc/users/<user_name>/<app_name>/local`:

```
## User context macro is to check response time of requests over 500
milli seconds [high_response_time]
definition = index=web_logs sourcetype=access_log response_time_
ms>=500
```

When you define a macro in the app directory, it sets a default configuration for the macro within the app. This means that all users who use the app will experience the same behavior when using the macro. In the app directory, you create another configuration file called `macros.conf`, located in `$SPLUNK_HOME/etc/apps/WebApp/local`. Within this file, you define the macro name, `high_response_time`:

```
## WebApp macro is to check the response time of requests over 1000
milliseconds
[high_response_time]
definition = index=web_logs sourcetype=access_log response_time_
ms>=1000
```

Having the same macro name defined in an app allows you to differentiate the behavior of the macro based on the context, and you can share the WebApp macro with the team responsible for investigating web performance.

The user-level configuration file in the user directory overrides the app-level configuration file in the app directory, according to search-time precedence. This allows users to personalize the behavior of the `high_response_time` macro to their specific needs. In the shared app context, the macro's definition is adjusted to filter for access logs with response times above 1,000 milliseconds. This higher threshold helps your entire team focus on more critical performance concerns that warrant collective attention. The configuration at the app level establishes the default behavior for searches that utilize the macro within the context of that specific app.

> **Tip**
>
> When you run a search in Splunk, the search-time objects, such as macros and other search-related configurations, are applied based on the context of the app in which the search is executed.

Scenario 2

In this scenario, we will look at the same setting created in two different apps.

App names can contain lowercase letters, uppercase letters, numbers, spaces, and special characters. If the same setting exists in two different apps, then the order in which Splunk decides to execute them at runtime is decided by sorting the apps in reverse lexicographical order.

For example, in the case of the same macro that we looked at in scenario 1, `[high_response_time]` is defined and shared with users that have access to the two apps, `WebApp` and `ITOpsApp`.

In this case, if both `WebApp` and `ITOpsApp` have a macro with the same name `[high_response_time]`, the macro definition in the `WebApp` app would take precedence over the one in the `ITOpsApp` app. This precedence is determined by the **reverse lexicographical order** of the app names.

This means that when Splunk resolves the `[high_response_time]` macro, it will first look in the `WebApp` app, and if it finds a definition there, it will use that definition. If no definition is found in the `WebApp` app, then Splunk will look in the `ITOpsApp` app:

```
## ITOpsApp macro is to check response time of requests over 2000
milli seconds
[high_response_time]
definition = index=web_logs sourcetype=access_log response_time_
ms>=2000
```

If you are curious about lexicographical and reverse order, Splunk Lantern has a very good article that goes into more detail: `https://tinyurl.com/22hjs8am`.

To avoid the hassle of manually checking each setting in every configuration file to understand how Splunk determines precedence during runtime, Splunk provides helpful tools such as the `btool` command and REST API interfaces. These tools allow you to easily examine the configuration settings and their priorities.

However, before diving into those tools, let's first focus on understanding precedence during index time.

Index-time precedence

Index-time precedence refers to the configuration settings that apply to the system as a whole, rather than being specific to individual users. These settings operate at a global system level and are involved in activities such as indexing data and monitoring files during the input phase.

During index time, various configuration files come into play, including `indexes.conf`, `server.conf`, `serverclass.conf`, and `inputs.conf`. These files contain essential configurations that affect how indexing activities are performed, how servers are configured, and how data inputs are monitored.

Splunk evaluates index time configurations in the following order:

1. `$SPLUNK_HOME/etc/system/local`

2. `$SPLUNK_HOME/etc/apps/<app_name>/local`

3. `$SPLUNK_HOME/etc/apps/<app_name>/default`

4. `$SPLUNK_HOME/etc/system/default`

Let's consider an example of index time precedence.

Let's assume you want to define the input configuration to monitor log files in a specific app called `ApplicationLogs`. You can achieve this by configuring the `inputs.conf` file within the app's context.

In the `inputs.conf` file of the `ApplicationLogs` app, you define the following stanza to monitor log files:

```
# inputs.conf Unix-style monitor stanza
[monitor:///var/log/application/*.log]
sourcetype = application_logs
index = dev_app
disabled = false
```

In this example, the stanza specifies that Splunk should monitor all log files with a `.log` extension in the `/var/log/application` directory. The logs will be assigned the `application_logs` source type and indexed in the `dev_app` index. The `disabled` parameter is set to `false`, indicating that the monitoring of these logs is enabled.

To override the `disabled` attribute, you can create a local version of the `inputs.conf` file in the appropriate directory, which has higher precedence than the `ApplicationLogs` app directory. In this case, the best directory to use would be the `etc/system/local` directory, as it has higher precedence than the global (index time) context.

You can create a new `inputs.conf` file or modify an existing one. Add the following stanza to override the `disabled` attribute:

```
[monitor:///var/log/application/*.log]
disabled = true
```

In this example, we set the `disabled` attribute to `true` to override the previous setting and disable the monitoring of log files in `/var/log/application`.

By placing the modified `inputs.conf` file in the `etc/system/local` directory, Splunk will prioritize this local configuration over the `etc/apps/ApplicationLogs` directory. This allows you to override specific settings such as `disabled`, providing flexibility to customize the required stanza behavior without affecting the rest of the system-wide configuration.

There is a special case in the `etc/apps/<app_name>` scenario. For example, if a setting/attribute is defined under the `[default]` global stanza present in two separate apps called `AppAlpha` and `AppBeta`, then Splunk evaluates the precedence in **lexicographical order**. This implies that preferences in the `AppAlpha` settings take precedence over those in the `AppBeta` directory.

In lexicographical order, numbers are prioritized over uppercase letters, and the least priority is given to lowercase letters. *Figure 6.2* is a good example of how Splunk handles apps with different names when merging configurations.

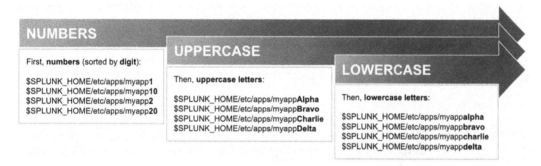

Figure 6.2: The lexicographical order of app directories in Splunk

In the upcoming section, we will explore the `btool` command to examine the configuration files and their settings.

Troubleshooting conf files using the btool command

As we have seen, conf files can be placed in more than one place, and in some cases, the default conf setting might need to undergo changes, which an administrator or Splunk user will create in `/local` directories, either in an `etc/apps/<app_name>` or `users/<user_name>/<app_name>` directory.

As more conf files are added to or updated, it becomes very difficult to track the changes to troubleshoot Splunk issues. In this situation, the `btool` CLI command is very helpful for seeing merged versions of conf files that exist on disk. Let's look at the syntax of the `btool` command. It must be issued from the `$SPLUNK_HOME/bin` directory:

```
./splunk btool  <conf_file_prefix> [list|layer|add|delete]  --debug
--app=<app_name> --user=<user_name>
```

Let's understand the components:

- --debug: This option shows the absolute path of the conf file's location in the system, app, or users directory. Similar to the --debug option, there are many useful options available. To view the full list, issue the following command – $SPLUNK_HOME/bin/splunk help btool.

- --user: This option displays the configuration data visible to the given user.

- --app: This option displays the configuration data visible from the given app.

- <conf_file_prefix>: This refers to the prefix used in the names of various configuration files – for example, props, inputs, indexes, and transforms.

Let's see a btool command example for the indexes.conf file with the --debug option:

```
./splunk btool indexes list --debug
```

In the given example, the btool command was used to *list* the *indexes* in *debug* mode. We will get the following output – as you can see in the following figure, the indexes list --debug option shows the absolute path of the conf file and each setting associated with the index name _internal.

```
C:\Program Files\Splunk\bin>splunk.exe btool indexes list --debug  findstr _internal
C:\Program Files\Splunk\etc\system\default\indexes.conf [_internal]
C:\Program Files\Splunk\etc\system\default\indexes.conf coldPath = $SPLUNK_DB\_internaldb\colddb
C:\Program Files\Splunk\etc\system\default\indexes.conf homePath = $SPLUNK_DB\_internaldb\db
C:\Program Files\Splunk\etc\system\default\indexes.conf thawedPath = $SPLUNK_DB\_internaldb\thaweddb
C:\Program Files\Splunk\etc\system\default\indexes.conf tstatsHomePath = volume:_splunk_summaries\_internal
db\datamodel_summary
```

Figure 6.3: The btool command result for indexes.conf

Another useful btool command is to find out whether there are any typos in a conf file. It usually runs every time a Splunk instance is started. The syntax is as follows:

```
./splunk btool check
```

Remember these facts about btool:

- The output generated by btool may not reflect the current state of the loaded memory. In the event that a change is made to a configuration file that requires a service restart, the btool report may reflect this change, even though it has not yet been applied in the active system state.

- btool is unsupported by Splunk; however, it is very powerful and useful. When dealing with Splunk support cases, they might ask for a Splunk diag report, which includes the btool report output.

- When viewing an inputs.conf file with btool, the [default] stanza settings are not shown in the output.

- You can execute a btool command without requiring the Splunk process to be in a running state.

Summary

Splunk configurations are stored in files with the .conf extension. They are called conf files for short. In this chapter, we started by understanding conf files and their directory order of precedence and ended by looking at troubleshooting using the btool command.

Let us go through the important items to remember while working with conf files. Conf files with the same stanza names and in multiple directories are merged during the runtime of Splunk execution. The precedence is decided based on the type of conf file and its context. The precedence types are index time and search time, and the order of consideration of app directories is reverse lexicographical for search time and lexicographical for index time files.

Search-time precedence gives the highest priority to the etc/users directory through etc/apps, and the lowest priority is the system directory. Index-time precedence gives the highest priority to the system/local directory through etc/apps and the lowest priority to the system/default directory.

Modifications made to configuration files through the filesystem won't take immediate effect. Instead, changes necessitate a debug/refresh, also known as a reload. This approach is applicable either in a standalone Splunk instance or on a search head instance.

As it is very hard to search through individual files across many directories to troubleshoot conf file issues, the btool command really helps to see the merged version of conf files. The splunk help btool CLI command presents the available options along with the corresponding syntax to utilize them. splunk btool check, by default, executes every time Splunk is started. It checks for typos and deviations from the specification of individual conf files.

> Tips
>
> Places to override the default settings are $SPLUNK_HOME/etc/system/local (to apply them system-wide), $SPLUNK_HOME/etc/apps/<app_name>/local, and $SPLUNK_HOME/etc/users/<user_name>/<app_name>/local.
>
> Package the configurations in an app and deploy them through **Deployment Server (DS)** to forwarders, to a **Search Head (SH)** cluster using an SHC deployer, and to an indexer cluster through a cluster manager. *Chapter 7* introduces the clustering of SHs and indexers.

Alrighty! Let's test what we have learned so far in the next section. And in the next chapter, we are going to learn about Splunk distributed search.

Self-assessment

You will be given 10 questions and answer to choose from. The pattern of the questions is the same as we discussed in *Chapter 1*. Post-assessment, refer to the sections related to anything you had difficulty answering questions about. Alternatively, you can refer to the Splunk documentation. All the best! Let's get started.

1. Which conf file is required to define parsing settings for line breaking?

    A. `outputs.conf`

    B. `line_breaker.conf`

    C. `props.conf`

    D. `server.conf`

2. Does creating a conf file directly on the filesystem reload a Splunk configuration automatically?

    A. Yes

    B. No

3. How do you reload conf file changes on a standalone Splunk instance? (Select all that apply):

    A. Add the `restart = true` setting at the end of the conf file.

    B. Use a debug/refresh endpoint.

    C. Restart the Splunk instance.

    D. Do nothing – Splunk reloads them automatically.

4. As a data admin, you are tasked with adjusting the `INDEXED_EXTRACTIONS` settings for an app named **MyApp**, located in the `/opt/splunkforwarder/etc/apps/MyApp` directory on a **Universal Forwarder** (UF). MyApp has, by default, included `props.conf` under the `default/` directory. What action will you take?

    A. Log in to UF and update `props.conf`, as the change is required on a single host only.

    B. Log in to UF, create a `/local` directory under MyApp, and create a `props.conf` file with the required settings.

    C. Do nothing – `props.conf` parsing does not work on UF.

    D. Do nothing – this is a system admin job, so inform the system admin.

5. Can you identify the location where conf files can be created or pre-exist on Splunk, where `$SPLUNK_HOME = /opt/splunk`? (Select all that apply):

    A. Location – `etc/apps/users/<user-name>/<app-name>/local/`

    B. Location – `etc/apps/<app-name>/default/`

C.   Location – `etc/apps/<app-name>/system/`

D.   Location – `etc/system/default/`

E.   Location – `etc/system/local/`

6.   Which configuration files are considered by Splunk for precedence during search time and index time? (Select all that apply):

A.   Search-time – `savedsearches.conf`, `macros.conf`, and `server.conf`

B.   Search-time – `savedsearches.conf`, `macro.conf`, and `tags.conf`

C.   Index-time – `inputs.conf`, `server.conf`, `indexes.conf`, and `savedsearches.conf`

D.   Index-time – `inputs.conf`, `serverclass.conf`, `indexes.conf`, and `server.conf`

7.   Using which CLI command do you troubleshoot conf file errors? (Select all that apply):

A.   `./splunk check errors`

B.   `./splunk btool check`

C.   `./splunk btool <conf-file-name> list --debug`

D.   `./splunk btool check-typos`

8.   User A has the power user role and has created an alert with private permissions under the search app. What location will it have been stored in on Splunk?

A.   `$SPLNUK_HOME/etc/users/local/`

B.   `$SPLNUK_HOME/etc/apps/search/users`

C.   `$SPLNUK_HOME/etc/users/User-A/search/local`

D.   `$SPLUNK_HOME/etc/users/search/local`

9.   User A has the power user role, has created an alert, and shared it with an app called `search`. What location will it have been stored in on Splunk?

A.   `$SPLNUK_HOME/etc/users/search/local`

B.   `$SPLNUK_HOME/etc/apps/search/local`

C.   `$SPLNUK_HOME/etc/apps/users/User-A/search/local/`

D.   `$SPLUNK_HOME/etc/apps/search/default/`

10. As a data admin, if you intend to create a `props.conf` file under the `$SPLUNK_HOME/etc/system/` directory on a forwarder that is set up to communicate with the DS, would you be able to accomplish this?

    A. Yes, it is possible. Create an app called `system` on DS.

    B. No, it is not possible.

Hopefully, the questions have helped you to understand the context of what we learned in this chapter. Let's review the answers in the next section.

Reviewing answers

1. *Option C* is the right answer. The `props.conf` file contains settings for line breaking, timestamp identification, line merging, and so on.

2. *No*, filesystem changes require either a restart of the Splunk instance or invoking the `/debug/refresh/` endpoint on a standalone Splunk instance that reloads most of the conf files.

3. *Options B* and *C* are the right answers.

4. *Option B* is right. It is always best practice to create a `/local` directory and modify the settings.

5. *Options A, B, D*, and *E* are the right locations for conf files creations. Users' private objects are stored in the `etc/users/<app-name>/local` directory. Objects shared in the app/global context are stored in the `etc/apps/<app-name>/[default|local]` directory. System-level settings are in the `/etc/system/[default|local]` directory.

6. *Options B* and *D* are the right answers.

7. *Options B* and *C* are the right answers. `./splunk btool` identifies conf files that are non-compliant with the specification. `./splunk btool <conf-file-prefix> list –debug` shows the merged conf file settings across various locations, based on the order of precedence with the location of conf files.

8. *Option C* is the right answer. Users' private object configurations are stored within their respective users' app directories. In cases where stanzas share the same name, the private configurations of users take precedence over any other location during search time.

9. *Option B* is the right answer. Objects shared by users are typically stored in the context of the app they are associated with. This includes objects shared within the app, across all apps (in the system), and globally. This default storage location is within the app itself. **Role-Based Access Control (RBAC)** of individual objects is controlled by `.meta` files (`default.meta`, `local.meta`) in the `metadata` directory of the app. For example, `App-A` contains `[default|local].meta` files in the `/etc/apps/App-A/metadata/` location.

10. *No*, it is not possible to change the conf file settings in the `etc/system/` directory on a forwarder through DS. The best practice is to package the settings in an app and deploy it.

7

Exploring Distributed Search

Splunk distributed search is all about the separation of search and indexing duties. In a standalone Splunk instance, there is no concept of distributed search because the duties of searching and indexing are handled by only one instance. If these duties are separated, there will be an opportunity to scale individual search heads and indexers to be more efficient and robust. In this chapter, you will learn about distributed search configuration and how knowledge bundles are replicated from search head to indexer. While distributed search is in action, there can always be scenarios where one or more instances might not perform as expected. In that case, we will examine the troubleshooting steps we can apply to overcome the issues or at least partially restore the search.

It is very important for system administrators to understand the various roles of Splunk instances in a distributed deployment and how search gets distributed to indexers. It is also important to be ready with mitigation strategies so that the search functionality can continue working.

In this chapter, we will cover the following topics:

- Understanding distributed search
- Search head and indexer clustering overview
- Configuring distributed search
- Understanding knowledge bundles

Understanding distributed search

In a Splunk distributed deployment, the main components involved in the distributed search are the **search head** and the **indexer/search peer**. A **distributed deployment** in Splunk connects multiple instances (search heads and indexers) to function as a unified system, enabling scalability, and efficient resource utilization.

The search head is responsible for presenting the Splunk interface to the user. It handles search management tasks and distributes search requests issued by the user to the indexers, which are also known as search peers. The search head coordinates the search process by sending search queries to the indexers and consolidates the results received from them. It then presents the aggregated results to the user.

On the other hand, the indexer is responsible for data indexing. It receives search requests from the search head and processes them by searching through the indexed data. The indexer returns the relevant results (based on the user's search query) back to the search head, which then consolidates and presents them to the user.

The **distributed search** feature in Splunk enables scalability and performance optimization by distributing the search workload across multiple indexers. This enables parallel processing of search queries, leading to faster search results and better resource utilization.

Let's explore some key facts about distributed search in Splunk in standalone, non-clustered, and clustered Splunk deployments:

- A standalone Splunk deployment does not use distributed search. In a standalone deployment, the entire search and indexing functionality is performed by a single instance of Splunk.

- To use the distributed search feature, Splunk instances (search head and indexer) do not need to be in a cluster. In other words, clustering is optional. A simple deployment with a single indexer and search head could be connected to use this feature. Refer to *Figure 7.1*.

- Distributed search comes into play when you have a small- to large-scale Splunk deployment with high data volumes and search requirements. It allows you to distribute the search workload across multiple Splunk instances, which may or may not be in a cluster.

- In a clustered environment, distributed search can be used in conjunction with clustering to enhance search capabilities. Each indexer in the cluster can participate in a distributed search, allowing search queries to be distributed across the indexers within the cluster.

For example, the following are some of the distributed deployment types that use the distributed search feature:

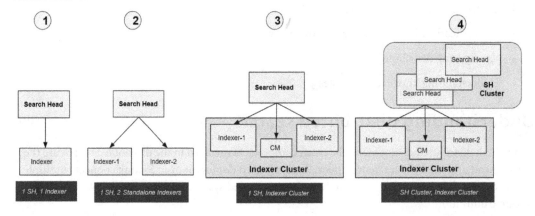

Figure 7.1: Distributed deployment types that use the distributed search feature

In *Figure 7.1*, we have the following:

- Deployment type 1 consists of a single standalone search head connected to a single indexer
- Deployment type 2 consists of a single standalone search head connected to two individual indexers
- Deployment type 3 consists of a single standalone search head connected to an indexer cluster
- Deployment type 4 consists of a search head cluster connected to an indexer cluster

In the next section, we are going to have an overview of search head clustering and indexer clustering.

Search head and indexer clustering overview

Clustering is a vast topic, and there is an administration course available from Splunk Education that covers it. In this section, you are going to learn about the essential topic of clustering in a distributed search context.

Clustering is the concept of grouping similar instance types to work toward a common goal by sharing the same configurations and objects, which allows resources to be highly available and resilient. Let's look at the two types of clustering in the following subsections.

Search head clustering

The search head in Splunk serves as the prime component responsible for managing user queries and coordinating the search process. Whether a search query is submitted through the Splunk user interface, the **Command-Line Interface** (**CLI**), or an API, the search head receives the query and takes charge of distributing the search across the indexers. It distributes the query to the appropriate indexers, collects and merges the results, and presents them to the user.

A search head cluster must contain at least three search head instances that share the same configurations, such as knowledge objects, apps, and artifacts, and share the job load. One of the instances in the cluster works as the captain and coordinates the cluster tasks. A separate instance is required to coordinate the deployment of apps to the cluster; that is the search head deployer. The key advantage of a search head cluster is high availability. Refer to *Figure 7.1* and look at deployment type 4.

To determine the number of search heads required, there is a range of factors to consider. For example, premium apps such as Splunk **Enterprise Security** (**ES**) and **Information Technology Service Intelligence** (**ITSI**) require a dedicated search head. A single search head with Splunk's recommended hardware specification supports 8-10 simultaneous searches, combining ad hoc and scheduled searches, and it can handle a small number of users. Refer to the *Splunk Validated Architectures* documentation, `https://tinyurl.com/3xtkppt9`, for guidance.

For more information, look at the following search head cluster architecture:

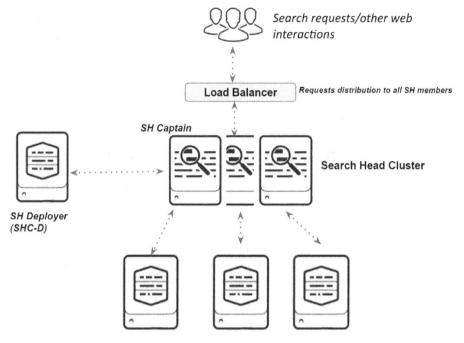

Figure 7.2: Search head cluster architecture

In *Figure 7.2*, there is a cluster consisting of three search head instances. These instances are fronted by a load balancer, which evenly distributes the load between the search head members in the cluster. Among the search head instances, one of them is elected to perform the role of the **search head captain**, which is responsible for coordinating various tasks within the cluster.

The search head captain instance handles tasks such as job scheduling, configuration replication, artifact replication, alert tracking, and knowledge bundle replication across the search head members. However, the captaincy role is not fixed to a specific instance. Instead, an ongoing election process occurs among the search head members to determine the winning member that will take over the captaincy role. This ensures that the responsibilities are dynamically assigned and can be redistributed as needed.

If it is necessary to manually transfer the captaincy from one instance to another, it is possible to initiate the transfer through a CLI command. This allows flexibility and control over the captaincy assignment within the cluster.

In addition to the cluster of search head instances and the captaincy functionality, the provided setup includes a separate instance called the **search head deployer**. The search head deployer serves as a centralized hub for deployment apps that are pushed out to the search head cluster when needed by administrators.

The search head deployer instance contains the deployment apps, which are configurations, settings, and customizations that need to be distributed across the search head cluster. These apps may include saved searches, dashboards, reports, field extractions, and other objects that enhance the functionality and customization of the search head instances.

Administrators can make changes to the deployment apps on the search head deployer and then push those changes to the search head cluster. This centralized approach simplifies the management and distribution of configurations across multiple search head instances, ensuring consistency in deploying and maintaining apps.

Indexer clustering

An indexer cluster contains a group of Splunk Enterprise instances that are configured to work as indexers (search peers) and they all share the same index configurations and apps. A separate instance is required to manage the indexer cluster activities; that is called a **cluster manager**.

An indexer cluster is required to meet the demands of high availability, fault tolerance, and scalability when dealing with substantial data volumes and concurrent search queries. By utilizing an indexer cluster, organizations can guarantee uninterrupted data access, mitigate the effects of indexer failure, and achieve optimized data indexing and search performance.

Indexer clusters provide several key benefits:

- **Data availability**: By distributing and replicating data across multiple indexers, an indexer cluster ensures that data remains accessible even if one indexer goes down or becomes unavailable.

- **Data recovery**: In the event of an indexer failure, the cluster retains copies of indexed data on other indexers. This redundancy enables automatic retrieval of the relevant data from the remaining indexers when a search request is issued.

- **Data fidelity**: Indexer clusters maintain data fidelity by replicating and distributing data across multiple indexers. This replication strategy safeguards data integrity and consistency, reducing the risk of data loss and ensuring that indexed data remains accurate and reliable.

Let's understand how searching works in the context of indexers. When a search query is issued by a user, the search head distributes the query to the indexers in the cluster. Each indexer performs a local search on its searchable copy of the data that resides locally in its storage. The search head then aggregates and merges the results from all indexers to generate the final search result set.

The presence or absence of index files within the data determines whether it is considered a searchable copy or non-searchable data:

- **Searchable copies**: Searchable copies consist of both raw data and its associated index files. These index files play a crucial role in enhancing search performance. When a search query is executed, Splunk uses the index files to quickly locate and retrieve the relevant data, resulting in faster search results. Searchable copies are the primary source for efficient search operations in Splunk.

- **Non-searchable data**: Non-searchable data refers to data that lacks index files. It typically includes only the raw data without any associated indexing. Searching directly within non-searchable data is slow and inefficient since Splunk does not have the index files to optimize the search process.

The **Replication Factor** (**RF**) on the indexer cluster determines the number of data copies to be replicated across the cluster, and the number of indexers in a cluster must be at least equal to the RF. For example, setting `replication_factor = 3` requires at least three indexers in a cluster along with a manager node. Each node maintains one copy, which equates to three data copies.

In addition to the RF, another important configuration in an indexer cluster is the **Search Factor** (**SF**), which is set on the manager node. The SF determines the number of searchable copies that should be available for the indexed data. The SF value must be equal to or less than the RF setting. The SF setting ensures that an adequate number of searchable copies are maintained across the cluster, enabling reliable search operations in Splunk.

If you look at *Figure 7.2*, the search peers at the very bottom are not in a cluster and they are working independently. They all maintain their own data copies without any replication, and if any of the peers are unavailable, there won't be a chance to search the data on the peer that went offline. Similarly, if one of them crashes, data recovery might not be even possible unless there was a reliable backup solution in place, and it is a tedious task to maintain a backup at the storage level as well.

So far, what we have seen is a single-site cluster running in one data center. What if you need to have *multiple indexer clusters* in more than one data center for the purposes of disaster recovery? That is possible through the **multisite indexer clustering** feature of Splunk Enterprise.

Consider a scenario where we have two data center locations with optimized network latency and efficient storage access. Within this setup, the search peers extend across these two data centers, comprising a single manager node. Additionally, each data center is equipped with a dedicated search head. The configuration involves **site RF** and **site SF** settings to guarantee data replication between the sites. This results in the creation and maintenance of searchable copies across the indexers situated in the two distinct data center sites. We learned about multisite clustering in the *Distributed clustered deployment and SHC – multisite* section of *Chapter 1, Getting Started with the Splunk Enterprise Certified Admin Exam*.

> **Tip**
>
> Splunk recommends disabling web servers on indexers if there is no requirement to use the Splunk Web user interface. To disable web servers, issue the `./splunk disable webserver` CLI command on indexers.

Configuring distributed search

The distributed search feature requires configuration to establish connectivity from the search head to indexers or search peers. From a standalone search head and indexer architecture to a large-scale multisite clustering setup, all implementations make use of the distributed search feature. Refer to the deployment type in *Figure 7.2*, which has three search heads in a cluster, enabling distributed search to interact with three independent indexers. Indexers receive data from various sources, such as universal forwarders, syslog inputs, and technology add-ons for indexing. Here are a few essential points to understand about distributed search:

- Search heads are preconfigured to send queries to search peers upon user request.

- Search heads consolidate the results received from peers that participate in the search.

- Search heads present the results to the user.

- A continuous background process called knowledge bundle replication runs (you'll learn about this later in this chapter) on the search head captain, which keeps sharing the information with search peers. In a non-cluster search head environment, a standalone search head shares the bundle with all the search peers it is connected to.

- Splunk recommends dedicating a separate Splunk Enterprise instance to each role, such as a deployment server for forwarder management, a search head deployer for the search head cluster, a license manager for licensing, and a monitoring console for overall deployment monitoring. However, in underutilized deployments, management components can be colocated according to the matrix provided at `https://tinyurl.com/26f9n5zf`.

> **Tip**
>
> In a distributed deployment, the best practice is to avoid indexing search head internal logs locally and enable forwarding on the search head. So, the internal logs and summary index data will be stored centrally on indexers.

Distributed search can be configured through Splunk Web, the CLI, or by directly modifying the `server.conf` and `distsearch.conf` files. In the next section, we will learn about adding indexers to a search head via the CLI and Splunk Web approaches.

The Splunk CLI

In the two upcoming subsections, we will use Splunk CLI commands to establish a distributed search configuration. This will include configuring distributed search from a search head to independent/non-clustered indexers, as well as setting up distributed search from a search head to clustered indexers.

Adding non-clustered indexers

Let's look at using the Splunk CLI to configure distributed search. The following command adds the search peer information to the `distsearch.conf` file in the `$SPLUNK_HOME/etc/system/local` directory. This approach is suitable for standalone non-clustered indexers, and the command should be executed on the search head.

Change the directory to `$SPLUNK_HOME/bin` and run this:

```
./splunk add search-server <scheme>://<host>:<port> -auth
<user>:<password> -remoteUsername <peer-user> -remotePassword <peer-
password>
```

The same command should be repeated for every indexer/search peer.

This command contains the following elements:

- `<user>:<password>` is the search head admin-level credentials
- The `<peer-user>` and `<peer-password>` credentials should correspond to an admin-level user on the target indexer, or at the very least, a user with the necessary `edit_roles` permissions
- `<scheme>://<host>:<port>` is the search peer details

For example, run the following command to add an indexer peer host called `indexer-host1` on the `$SPLUNK_HOME/bin` search head:

```
./splunk add search-server https://indexer-host1:8089 -auth
admin:password -remoteUsername admin -remotePassword password-of-
remote-peer-goes-here
```

Let's look at adding clustered indexers/search peers in the next section.

Adding clustered indexers

Let's see what information we need to add an indexer cluster to a search head. When an indexer cluster is configured, the Splunk administrator usually sets a secret/pass4SymmKey/security key. The pass4SymmKey has the same value across all indexer peers. A pass4SymmKey enables secure authentication between search heads and indexers. In addition to a key, we need the cluster manager's URI and management port details.

To add a single-site clustered indexer to a search head, issue the following command in the `$SPLUNK_` `HOME/bin` directory on the search head. Behind-the-scenes configuration settings will be added to the `server.conf` file in the `$SPLUNK_HOME/etc/system/local` directory:

```
./splunk add cluster-manager <scheme>://<host>:<port> -secret
<indexers-pass4symmkey> -multisite false
```

This command must be repeated for every cluster manager.

Let's look at the elements of this command:

- `<scheme>://<cluster-manager-servername>:<management-port>` is the cluster manager details
- `<indexer-pass4symmkey>` is the pass4SymmKey/secret/security key of the indexer cluster

As an illustration, assuming you've previously executed a command to add the cluster manager address `SplunkManager01.example.com`, the `server.conf` file now displays the following content:

```
[clustering]
mode = searchhead
manager_uri = https://SplunkManager01.example.com:8089
pass4SymmKey= encrypted-secreteast
```

For further options, such as listing, editing, or removing a cluster manager, you can use the `help` command. For instance, execute `./splunk help cluster-manager` in the `/bin` directory of `$SPLUNK_HOME`.

Similarly, if you wish to learn how to edit and/or remove `search-server`, you can employ the `help` command as follows: `./splunk help search-server`. This should be done within the `/bin` directory of `$SPLUNK_HOME`.

We have now learned how to include both clustered and non-clustered indexers in a search head for the purpose of configuring distributed search. Moving forward, the subsequent section will guide us through the process of configuring distributed search using the Splunk Web interface.

Splunk Web

Using Splunk Web, we can create the same configuration that we did in the previous section with the CLI. Let's look at the menu options in *Figure 7.3*.

Log in to Splunk Web as an admin user, navigate to **Settings** in the top-right corner, and click on it. Then, you will find the following menu items under **DISTRIBUTED ENVIRONMENT**:

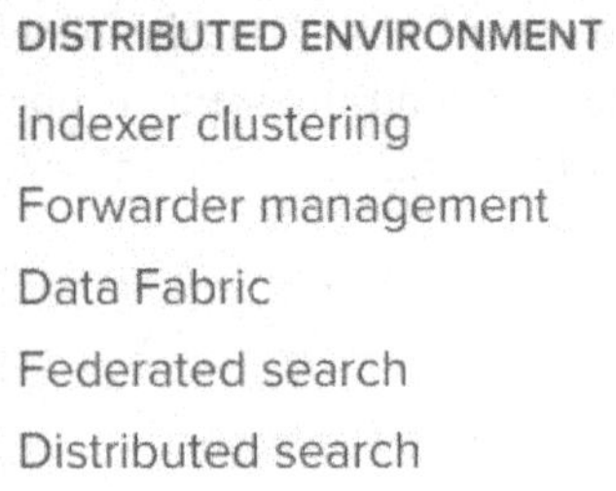

DISTRIBUTED ENVIRONMENT

Indexer clustering

Forwarder management

Data Fabric

Federated search

Distributed search

Figure 7.3: Distributed search web menu

In the next section, we are going to add non-clustered indexers to the search head.

Adding non-clustered indexers

Figure 7.3 has two menu items that are related to distributed search, which are **Indexer clustering** and **Distributed search**. Let's click on **Distributed search** and, on the next page, click on + **Add New (search peers)**, which will open the following form:

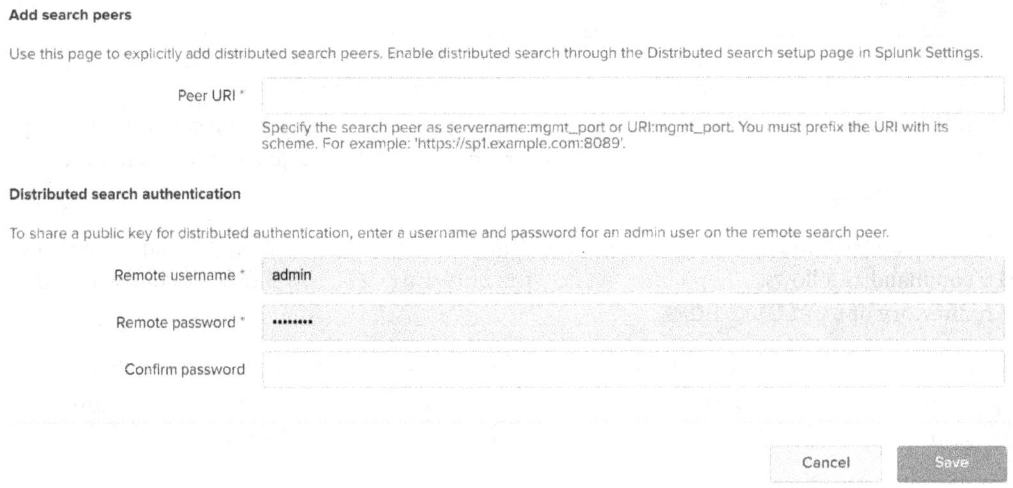

Add search peers

Use this page to explicitly add distributed search peers. Enable distributed search through the Distributed search setup page in Splunk Settings.

Peer URI *

Specify the search peer as servername:mgmt_port or URI:mgmt_port. You must prefix the URI with its scheme. For example: 'https://sp1.example.com:8089'.

Distributed search authentication

To share a public key for distributed authentication, enter a username and password for an admin user on the remote search peer.

Remote username *   admin

Remote password *   ••••••••

Confirm password

Cancel   Save

Figure 7.4: Distributed search – Add search peers

Figure 7.4 has the following text fields: **Peer URI** (peer-URI:mgmt_port format), **Remote username**, and **Remote password**, in which you can manually enter the details of the indexer that needs to be added to the search head. Populate these fields and click **Save**. Repeat the same steps as shown in *Figure 7.3* and *Figure 7.4* for every search peer to add them to the search head.

In the next section, we are going to add clustered indexers to the search head.

Adding clustered indexers

In Splunk Web, click on the **Settings** menu and click on **Indexer clustering**, as shown in *Figure 7.3*, which will open a new page that has an **Enable Indexer Clustering** button. Click on the button and select the **Search head node** radio button, as shown in *Figure 7.5*.

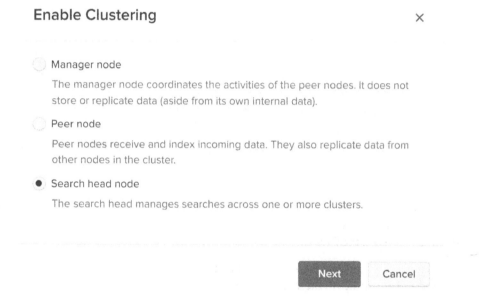

Figure 7.5: Distributed search – Enable Clustering on the search head

Figure 7.5 shows the options to enable indexer clustering on the search head. Finally, click on the **Next** button. A form will open, as shown in *Figure 7.6*.

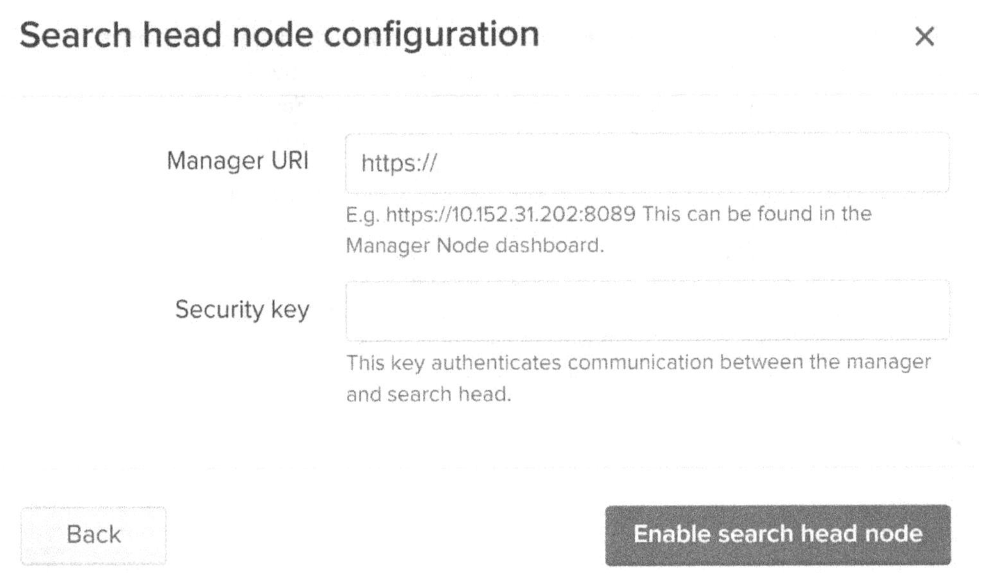

Search head node configuration   ✕

Manager URI
> https://

E.g. https://10.152.31.202:8089 This can be found in the
Manager Node dashboard.

Security key

This key authenticates communication between the manager
and search head.

[Back]   **Enable search head node**

Figure 7.6: Distributed search – adding a cluster manager

Figure 7.6 shows the final form for adding a cluster manager to the search head. Complete the form by filling in the cluster manager URI and cluster pass4Symmkey/secret key and click on **Enable search head node**.

Great! Distributed search is now ready to issue search requests on the search head. Let's revise what we have learned. We started by discussing distributed search, followed by adding non-clustered and clustered indexers through the CLI and Splunk Web approaches.

In the next section, we will explore the knowledge bundle replication process, which originates from the search head and propagates to the associated search peers.

Understanding knowledge bundles

A **knowledge bundle** is an archived file containing knowledge objects, and it is distributed by the search head to the search peers to process distributed searches. Knowledge objects may be related to field extractions, saved searches, reports, alerts macros, event types, lookups, and user authorization, among other things. Knowledge objects within Splunk can be categorized as either private to an individual user, shared across specific applications, or even global (accessible to all users). Moreover, access to these objects can be restricted to particular user roles, adding an extra layer of control and security.

For a more comprehensive explanation of managing object permissions, I suggest referring to the official Splunk documentation at the following link: `https://tinyurl.com/ytr2ckaw`. This resource delves into a wide range of scenarios that extend beyond the scope of the exam.

Search peers receiving the knowledge bundle participate in the distributed search process, which means peers apply applicable knowledge objects to the data before being returned to the search head. When a user issues a search request, the peer determines the applicable knowledge object based on the user authorization information received inside the knowledge bundle.

Knowledge bundle replication

Replication of knowledge bundles to search peers is a continuous process in the background of the search head or when a search is initiated. Replication in the search head cluster environment is the responsibility of the elected captain. In standalone search heads, it does the replication to all peers. Let's look at the key facts about replication:

- A replication policy is a setting on the search head that determines how to replicate the knowledge bundle to the peers. By default, it is set to `classic`, in which the search head distributes the bundle to all the peers. Another prominent policy is `cascading`, which is suitable for large-scale deployments for faster bundle replication. The setting is configured in the `distsearch.conf` file:

  ```
  [replicationSettings]
  replicationPolicy = [classic | cascading | rfs | mounted]
  ```

- The bundle files are usually full bundles with the `.bundle` extension and delta bundles with the `.delta` extension. Full bundles contain every knowledge object configured to replicate, and delta bundles only contain recently changed knowledge objects. During the first replication, the search head transfers the full bundle to its peers. Subsequent replications involve only the smaller delta bundle, which captures the incremental changes.

- The search head stores the bundle locally at `$SPLUNK_HOME/var/run`, and search peers store the received bundles at `$SPLUNK_HOME/var/run/searchpeers`. You can open bundle files and look at the contents using the `tar` command in Linux.

- Knowledge bundles contain files from `$SPLUNK_HOME/etc/system`, `$SPLUNK_HOME/etc/apps`, and `$SPLUNK_HOME/etc/users`.

- The bundle size on the search head is set by default to 2 GB via `maxBundleSize` in the `distsearch.conf` file, and if the size of the bundle exceeds the limit, then bundle replication is suspended until either the `maxBundleSize` limit is increased or the bundle size is reduced. On the receiving side, `max_content_length` in `server.conf` must be set with the same value as `maxBundleSize`.

- To reduce the overall bundle size, apps with large binaries, lookups, and other configurations that are not required on search peers for search processing can be added to the replication deny list using `[replicationDenylist]` in `distsearch.conf`.

- To look at the complete list of allowed configurations that are included in the knowledge bundle, refer to [replicationAllowlist] and [replicationSettings:refineConf] inside the distsearch.conf file of the search head. To check the bundle replication status, you can issue the splunk show bundle-replication-status command on the search head, or you can use the monitoring console, which has a dashboard under its search menu.

In cases where the common knowledge bundle version cannot be replicated across all search peers, it might lead to a temporary search interruption. The search process may pause for a few seconds as it attempts to synchronize the knowledge bundle across peers. If establishing a common bundle proves unsuccessful, the search process will eventually give up and resume without achieving synchronization. In that case, the Splunk administrator will take the necessary action by identifying the slow search peer and quarantining it on the search head to prevent the peer from participating in future searches. After **quarantine**, the search peer ensures that already running searches are completed. The quarantine action suspends knowledge bundle replication to only that peer.

Quarantine can be achieved through Splunk Web by clicking on **Settings | Distributed Search | Search peers**. You can also do it through the CLI. After successful restoration, **unquarantine** the peer so it can participate in future searches.

Here are the CLI quarantine and unquarantine commands:

```
./splunk edit search-server -auth <user>:<password> <host>:<port>
-action [quarantine|unquarantine]
```

Let's examine this in more detail:

- <user>:<password> is the search head's admin-level credentials
- <host>:<port> is the search peer's host and management port
- The -action values are quarantine or unquarantine

Let's conclude; a knowledge bundle is an archive file that is shared with search peers by the search head through a background replication process. A knowledge bundle contains knowledge objects that are applied by search peers to search results before being returned to the search head. The bundle size limit is applicable by default, and to reduce its size, certain knowledge objects can be added to the replicationDenyList setting on the search head. To suspend bundle replication to a particular peer, you can quarantine that search peer.

Let's summarize what we have learned in the next section.

Summary

In this chapter, we learned that distributed search separates the duties of search heads and indexers. Search heads accept search requests from users and distribute them to indexers that are preconfigured. Search results returned from indexers will be consolidated by the search head and presented to the user.

We looked at the advantages of search head clusters compared to standalone instances and indexer cluster architectures. The search head captain plays a vital role in distributing knowledge bundles, scheduling searches, artifact proxying, and more. A dedicated cluster manager in the indexer cluster replicates data copies and ensures the search factor is met.

After this, we went through the ways to configure distributed search on a search head using the Splunk Web and CLI approaches. After configuring distributed search, we examined knowledge bundles, what they contain, and how to minimize their size through the max bundle size and deny list settings, along with ways to monitor bundle replication activity to peers. At the very end, we learned how to troubleshoot bundle replication issues and how to quarantine and unquarantine a peer.

With this chapter, you have reached the end of *Part 1*, covering Splunk system administration topics. The next chapter begins *Part 2* of the book, which will cover Splunk data administration topics. To begin with, the upcoming chapter will dive into getting data into Splunk.

In the next section, there are some questions to test your knowledge about Splunk distributed search concepts.

Self-assessment

Answers are provided at the end of this section. If you are having trouble answering questions in certain areas, make sure you revisit those sections, and for more detailed information, you can refer to the Splunk documentation. Let's begin the questions:

1. What is a search peer? (Select all that apply.)

    A. Something that accepts search requests from the search head

    B. An indexer

    C. One of the search head members in a cluster

    D. All of the above

2. What is the role of a cluster manager? (Select all that apply.)

    A. To make sure the indexers are up and running all the time

    B. To configure the indexer cluster

    C. To replicate data copies across cluster members

    D. To make sure search copies are available

3. Is the search head captain component mandatory in a search head cluster?

    A. Yes

    B. No

4. You are assigned to a critical project where a search head cluster is configured to search multiple non-clustered indexers. Is it making use of the distributed search concept?

    A. Yes

    B. No

5. Identify the configuration files containing indexer details on a search head.

    A. `serverclass.conf`

    B. `server.conf`

    C. `distsearch.conf`

    D. `distributedpeers.conf`

6. What does a knowledge bundle contain? (Select all that apply.)

    A. Users, knowledge objects, and apps

    B. Knowledge objects such as event types, tags, and saved searches

    C. Splunk binary files

    D. Splunk internal logs about the search process

7. What is the default replication policy of knowledge bundles?

    A. Cascading

    B. RFS

    C. Classic

    D. Mounted

8. A knowledge bundle has grown so big that it exceeds `maxBundleSize`. You are a system administrator; what will you do?

    A. Increase `maxBundleSize`

    B. Delete the apps from the search head that are large

    C. Add files/apps that are not required to `replicationDenyList`

    D. Do nothing; the search head shares only the .delta file

9. Which configuration do you use to add non-clustered indexers to the search head?

   A. `server.conf`

   B. `distsearch.conf`

10. You have found a search peer that is slowing the search requests. What actions will you take?

    A. Shut down the search peer causing trouble

    B. Remove the search peer from distributed search

    C. Quarantine the search peer on the search head

    D. Do nothing; let the search peer recover

Reviewing answers

1. *Options A and B* are the right answers. A search peer is an indexer that participates in searches by accepting requests from the search head.

2. *Options C and D* are the right answers. A cluster manager ensures that raw data is copied to other members for data high availability. It tries to maintain the number of copies defined in the `replication_factor` setting. Similarly, it maintains the number of searchable copies defined by the search factor setting.

3. *Yes*. It is mandatory to have a search head captain in a cluster. The captain is chosen from the members through an election process.

4. *Yes*. The distributed search concept is applied to deployments where the search head and indexer roles are separated in a deployment. There is no dependency on clustering.

5. *Options B and C* are the right answers. Both the `server.conf` and `distsearch.conf` files contain indexer details in a distributed deployment; `distsearch.conf` is useful in non-clustered indexers, and `server.conf` is useful in clustered indexers.

6. *Options A and B* are the right answers. A knowledge bundle is an archive file that contains user knowledge objects and apps. Knowledge objects could be lookups, macros, saved searches, event types, and so on.

7. *Option C* is the right answer. Classic is the default replication policy for knowledge bundles.

8. *Options A and C* are the right answers. As an admin, you could increase the default `maxBundleSize` limit from 2 GB to an acceptable size. The other option is to identify the configurations or knowledge objects and apps that do not need to be shared with the peer and add them to `replicationDenyList`.

9. *Option B* is the right answer. The `distsearch.conf` file contains non-clustered indexers configurations on the search head.

10. *Option C* is the right answer. You can remove/shut down the peer; however, to investigate it, the troubled peer on the search head can be quarantined. After the search peer is working as expected, it can be unquarantined.

Part 2:
Splunk Data Administration

Part 2 of this *Splunk 9.x Enterprise Certified Admin Guide* is a deep dive into the world of Splunk data administration. It will equip you with the skills needed to efficiently bring data into Splunk, configure data inputs, perform parsing and transformation, extract meaningful fields, and enhance data with lookups. The self-assessment mock exam ensures that you can apply your knowledge effectively.

This part contains the following chapters:

- *Chapter 8, Getting Data In*
- *Chapter 9, Configuring Splunk Data Inputs*
- *Chapter 10, Data Parsing and Transformation*
- *Chapter 11, Field Extractions and Lookups*
- *Chapter 12, Self-Assessment Mock Exam*

8
Getting Data In

Welcome to *Part 2* of this book, dedicated to Splunk data administration. In the previous chapters, we focused on Splunk system administration, covering essential topics such as installation, license configuration, user management, index creation, distributed search configuration, Splunk forwarder management, and Splunk Validated Architectures. Now, let's imagine that you have a fully functional Splunk Enterprise instance with all the necessary system configurations in place. It is ready to receive and index data from various sources. In *Chapter 1*, we discussed the process of installing Splunk Enterprise in standalone mode. Please refer to the *Splunk installation – standalone* section for detailed instructions.

From this point onward, we will use the terms "data collection" and "data input" interchangeably throughout the chapters. Both terms refer to the process of gathering and retrieving data into Splunk for indexing and analysis.

In this chapter, we will learn about the fundamental data collection methods offered by Splunk. These methods are crucial for efficiently getting data into Splunk. Understanding these data collection methods is important because they allow you to gather data from various sources, such as log files, network data, APIs, and more. This versatility helps you collect data from your entire IT infrastructure, enabling effective monitoring, troubleshooting, and analysis. *Chapter 9, Configuring Splunk Data Inputs,* delves into these collection methods in detail.

By mastering these methods, you can centralize, index, and analyze data in Splunk, providing valuable insights and actionable intelligence for your organization.

In this chapter, we are going to cover the following topics:

- Understanding Splunk data inputs
- Understanding metadata fields
- Data indexing phases
- Splunk Web – Add Data feature

Understanding Splunk data inputs

Splunk Enterprise accepts any text data from a variety of sources such as computers, network devices, **Internet-of-Things** (**IoT**) devices, sensors, virtual machines, containers, databases, and so the list goes on. The source just needs to find the right data input type to ingest its text data into Splunk.

Data input types are a mechanism for forwarding data from the source to Splunk for indexing. Splunk offers the following five types of data inputs that work for most sources. In *Chapter 9*, *Splunk Data Inputs*, we'll go through input types in more detail:

- **File and directory monitoring**: Sources containing data in files and lists of files inside directories can be monitored for indexing. Here, files are monitored continuously as data is appended to the end of a file. Usually, a **Universal Forwarder** (**UF**) agent runs on the source system and monitors the files and directories. A UF offers another option to index files, called `Batch`, which is useful for indexing large files followed by automatic deletion or sending to a sinkhole.

- **Network data input**: The source can send data to a target listening network port that is ready to receive data. Both the **Transmission Control Protocol** (**TCP**) and **User Datagram Protocol** (**UDP**) are supported by Splunk.

- **Scripted input**: Here, a script is configured to run at intervals and produces output. Output data generated by the script will be indexed. Usually, a UF is used on source systems to run scripted inputs.

- **HTTP Event Collector** (**HEC**): There might be cases where a UF cannot be installed on the source, and instead the source sends data over APIs. HEC is the best bet for sending data over the Splunk REST API interface. This is often called *agentless data input*. In a distributed deployment, HEC is typically enabled on HFs and indexer components.

- **Windows technology add-on** (**TA**): These add-ons work with their respective technology to forward their data to Splunk. The Windows TA works for Windows hosts only. TAs are available for many other technologies besides Windows from the `https://splunkbase.splunk.com/` website.

In addition to standard data collection methods, there are alternative ways to index logs in Splunk. One option is to utilize third-party syslog collectors or enable syslog input on an indexer or intermediate **Heavy Forwarder** (**HF**). To configure data inputs, you can edit the `inputs.conf` file on the Splunk forwarder (UF/HF) or indexer.

Another convenient approach to uploading data and observing its appearance in Splunk is through the Splunk Web interface. Additionally, the Splunk **Command-Line Interface** (**CLI**) provides commands to configure data inputs. For a more comprehensive understanding, a later section of this chapter will explore the **Add Data** feature in Splunk Web in detail.

After data is forwarded using one of the data input methods mentioned earlier, it undergoes a parsing phase and eventually gets indexed. In order to effectively query the data using Splunk **Search Processing**

Language (SPL) commands, it must be associated with identifiable metadata fields assigned to each data event during the indexing process.

In the upcoming section, we will explore the concept of metadata fields in more detail. Following that, we will delve into the various phases of data indexing in a subsequent section.

Understanding metadata fields

The following list details the default metadata fields assigned by Splunk during the inputs phase. Note that *field names are always case-sensitive*. We will discuss the different phases that data goes through in the *Data indexing phases* section:

- `host` – This describes from which host, device, or machine the data originates.
- `source` – This represents the input source or origin of the indexed data.
- `sourcetype` – The type of machine data. For example, on Windows hosts, we may see `WinEventLogs` or `ActiveDirectory`.
- `index` – This allows us to provide an index name in the `inputs.conf` file; otherwise, `main` is used by default.
- `_time` – This records the time of the event in Unix-epoch format. Splunk tries to automatically detect this during the parsing phase, or alternatively the administrator can configure this through the `sourcetype` settings.

If the data input doesn't specify metadata fields in `inputs.conf`, then Splunk assigns the default metadata fields. For example, let's look at the following code snippet for monitoring a file on a Linux filesystem. Note that there are no metadata fields specified other than `sourcetype`:

```
# Example inputs.conf
[monitor:///var/log/httpd]
sourcetype = access_common
```

The following defaults will be used:

- `host` is set to set to "hostname of forwarder is running" by default
- `index` is set to `main` by default
- `source` is set to `/var/log/httpd` by default

In the provided code snippet for monitoring a file on a Linux filesystem, the `sourcetype` is explicitly set to `access_common` within the monitoring stanza.

Let's look at some more details on `sourcetype` in the following section.

Source types

In Splunk, a source type is used to specify how the data should be formatted during the indexing process. By defining a source type, Splunk knows how to handle and interpret the data, ensuring that it is properly indexed and made available for searching and analysis. Splunk Enterprise ships with default pre-trained source types and data administrators can define new custom types if required. The `sourcetype` field is assigned by Splunk to every data event that is indexed.

To find the default source types through the CLI, go to `$SPLUNK_HOME/etc/system/default` and find the `props.conf` file. The file contains the most widely used `sourcetype` definitions.

They can also be found through Splunk Web. After logging in, navigate to the **Settings** menu in the top-right corner and click on **Source types**. This will show you the most popular ones and the **New Source Type** button lets you create your own as well. Clicking on individual source types opens an **Edit Source Type** form with **App**, **Category**, **Timestamp extraction**, and **Event Breaks** options. The **Advanced** tab shows the `<Name/Key> = <value>` pairs for the settings as configured in the `props.conf` specification. You can refer to the full specification at `https://tinyurl.com/mra548tt`.

Some popular pre-trained source types are `_json`, `access_combined`, `csv`, `linux_secure`, `linux_audit`, and `cisco:asa`.

The following figure shows the **Edit Source Type** form for the `access_combined` source type.

Edit Source Type: access_combined	✕

Description	National Center for Supercomputing Applications (NCSA) combined format HTTP web ser
Destination app	system ▾
Category	Web ▾
Indexed extractions ?	none ▾

Event Breaks   Timestamp   **Advanced**

Name	Value	
CHARSET	AUTO ▾	✕
DATETIME_CONFIG		✕
REPORT-access	access-extractions	✕
SHOULD_LINEMERGE	false	✕

Figure 8.1: The Edit Source Type view in Splunk Web

Now that we have a clear understanding of data input types and metadata fields in Splunk, let's delve into the phases that data undergoes before it is indexed.

Data indexing phases

As data moves from the source machine to the Splunk indexer, it primarily goes through three phases. Note that this is a simplified version of the system for learning purposes; there are in reality more queues involved as data moves through the processing pipelines.

Splunk indexing is the process of ingesting and storing data in the Splunk platform for later analysis and searching.

As a data administrator, it's crucial to understand the different phases involved in Splunk indexing. The core competency of Splunk lies in its ability to search and analyze large volumes of data in real time, providing valuable insights. The indexing phase involves ingesting and storing data in the Splunk platform while the parsing phase involves the extraction of relevant information using pre-built or custom-defined rules. During the indexing process in Splunk, there are several significant activities that take place. These include data manipulation, the creation of indexed fields, and the calculation of licensing. By understanding these phases, data administrators can better leverage the capabilities of Splunk for effective data analysis and problem-solving.

The following figure shows the three phases:

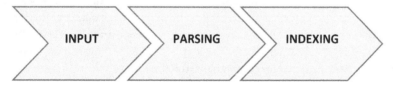

Figure 8.2: Phases of the data indexing process

Input

The input phase serves as the initial stage in the data indexing process, where data is brought in for indexing. On the source machine, the UF agent is commonly used to monitor and forward files, along with other data input types discussed earlier. Additionally, Splunk offers technology-specific add-ons from `splunkbase.com` to facilitate data collection for indexing.

During this phase, the configuration settings on the forwarder are typically specified in the `inputs.conf` file. Metadata fields such as default `host`, `source`, `sourcetype`, and `index` can be set here, although they can be overridden in subsequent phases as needed. The data collected in this phase is then passed on to the parsing phase for further processing.

Parsing

Event processing in Splunk occurs in two phases, parsing and indexing. During the parsing phases, all data goes through the parsing process to break it into individual events. This phase is crucial as it involves assigning timestamps to events, breaking events, and setting the time zone based on the source type definition.

The parsing phase relies on configuration settings in files such as `props.conf` and optionally `transforms.conf` to override default parsing settings and create new source type definitions. These configurations allow for customization and fine-tuning of the parsing process. Additionally, metadata field values from the input phase, such as `host`, `sourcetype`, and `index`, can be overridden or modified during this stage.

Another important aspect of the parsing phase is the ability to configure `nullQueue`, a special queue that can be used to drop unwanted events from being indexed, effectively filtering them out.

In short, the parsing phase in Splunk is responsible for breaking data into events, assigning timestamps, setting time zones, and providing flexibility for data manipulation and event filtering.

Let's look at an example scenario. By default, event breaking in Splunk occurs for every new line in the data. However, if you wish to modify this behavior, you can customize the `LINE_BREAKER` setting. This allows you to define your own pattern for event breaking based on specific requirements.

Another scenario we can consider is extracting timestamps. By default, Splunk searches up to a maximum of 128 characters of an event. If the timestamp in an event extends beyond this character limit, you can adjust the parameters of the extraction process using the `TIME_PREFIX` and `MAX_TIMESTAMP_LOOKAHEAD` settings in the `props.conf` file. This allows you to increase the range of characters to be considered for timestamp extraction.

Customizing these settings in the `props.conf` file provides flexibility in event breaking and timestamp extraction, enabling data administrators to adapt the parsing process to suit the unique structure and format of the data in question.

Chapter 10, Data Parsing and Transformation, examines the data manipulation conf settings and source type creation process in more detail.

After the data manipulation, event breaking, and timestamp extraction are complete, the data is handed over to the next phase: indexing.

Indexing

Indexing is the last phase of this process, where the data is written to disk. During this phase, index files are created, data is segmented for searching, the license is calculated based on the size of data to be indexed, and the raw data is compressed before being written to disk.

If you would like to dive into a detailed overview of these phases, along with queues and pipelines, then check out `https://tinyurl.com/bdzbcetp`. Note that this is completely optional for the certification exam.

We have now examined the main data ingestion phases and metadata fields. Next, we will explore the **Add Data** feature of Splunk Enterprise available in Splunk Web.

Splunk Web – Add Data feature

The **Add Data** feature is available in Splunk Web under the **Settings** menu. Simply open this menu and click on **Add Data**, as shown in *Figure 8.3*.

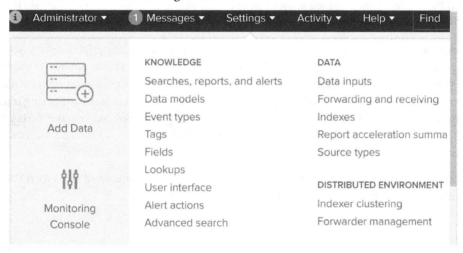

Figure 8.3: Splunk Web – Add Data

Clicking **Add Data** opens a new web page offering three ways to get the data into Splunk, as shown in *Figure 8.4*.

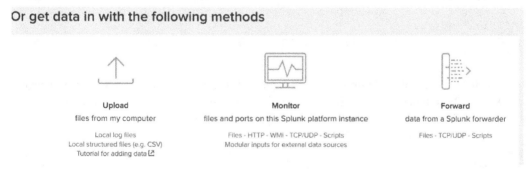

Figure 8.4: Methods for getting data into Splunk

As you can see, the first method is a one-time **Upload** method used to add data to Splunk in one go. The maximum file size for upload is 500 MB.

The second **Monitor** option gets data by monitoring the files and ports on the machine Splunk is installed on. This method allows the use of a number of data input options, such as turning the instance into an HEC receiver, enabling network ports to listen for TCP/UDP traffic, and configuring a script to pull data payloads from an API at defined intervals.

The third option in the preceding figure is **Forward**, which is specifically designed for a single-instance deployment of Splunk Enterprise where the Splunk instance serves as both the indexer and the deployment server. When selecting the **Forward** option, you can choose from the list of already connected Forwarder clients. The subsequent steps enable you to configure the specific data inputs that can be set up on the Forwarder.

Let's quickly look at the **Upload** option. The **Upload** option is a useful feature for quickly testing a data sample and identifying event line breaks. It allows you to upload a data file to Splunk for analysis, enabling you to identify how events are being broken down within the data. This can be particularly helpful in troubleshooting and understanding the structure of your data before configuring inputs or parsing settings in Splunk. You can also choose from suitable pre-trained source types during this process and can customize the source type live by altering the `<Key> = <value>` settings and observing the outcome.

For demonstration purposes, I am using the following access log information saved to my local system with the name `web-access.log`. The file's contents are as follows:

```
127.0.0.1 - - [23/Apr/2023:10:28:12 -0400] "GET /index.html HTTP/1.1"
200 1240
127.0.0.1 - - [23/Apr/2023:10:28:13 -0400] "GET /styles.css HTTP/1.1"
200 876
127.0.0.1 - - [23/Apr/2023:10:28:15 -0400] "GET /scripts.js HTTP/1.1"
200 432
192.168.0.1 - - [23/Apr/2023:10:28:17 -0400] "GET /images/logo.png
HTTP/1.1" 200 34012
192.168.0.1 - - [23/Apr/2023:10:28:19 -0400] "GET /about.html
HTTP/1.1" 404 345
127.0.0.1 - - [23/Apr/2023:10:28:21 -0400] "GET /contact.html
HTTP/1.1" 200 2089
```

Now, let's click on the **Upload** button shown in *Figure 8.4*. This will open a **Select Source** web page with a **Select File** button. Click on the button and upload the `web-access.log` file from your local system. You will then be taken to the **Set Source Type** page shown in *Figure 8.5*.

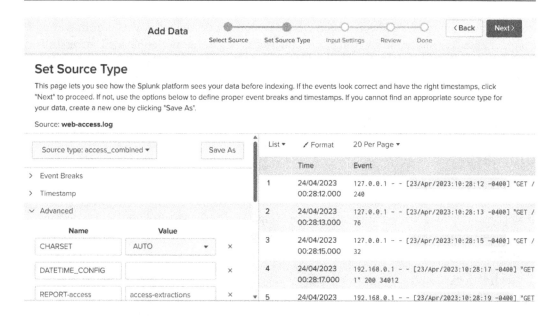

Figure 8.5: Add Data – Set Source Type

Figure 8.5 has three main sections to look at.

On the left-hand side, you can see that I have selected **Source Type: access_combined** . This is a pre-trained source type perfectly suited to accessing logs' structure. On the right-hand side, a list of events is presented with their timestamps detected. To observe the source type configuration settings, expand the **Advanced** section, which shows CHARSET, TIME_PREFIX, LINE_BREAKER, and so on, and their corresponding values.

Clicking the **Next** button will take you to **Input Settings**. This page is where you can set the host and index metadata fields or leave them at their defaults. Proceed and click the **Review** button to view a summary of the metadata fields' information (see *Figure 8.6*). If you are happy with the details, then click **Submit**; otherwise, click **Back** and edit the necessary sections.

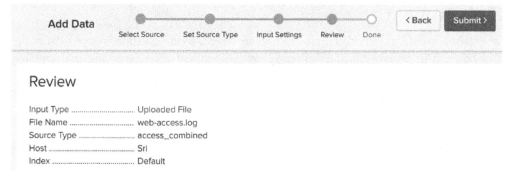

Figure 8.6: Add Data – Review stage

If everything has gone well, clicking the **Submit** button will index the uploaded `web-access.log` file on your Splunk instance and you will be presented with the **File has been uploaded successfully** message on the following page. This page also offers a lot of other options, including **Start searching**, **Extract Fields**, and **Add more data**. Clicking the **Start searching** button will open the Search app and show the uploaded `web-access.log` file events.

Let's conclude this chapter and review a summary of what we've learned so far.

Summary

This first chapter of *Part 2* of the book aimed to get you started with Splunk data administration. We began with the introduction of data input types, including the file-based, network, agentless (HEC), and script-based options. There is also a special type of input that can be installed through TAs available from `https://splunkbase.com`. We also understood that these inputs are configured either by creating an `inputs.conf` file or through the Splunk CLI.

Afterward, we looked at the default metadata fields assigned by Splunk, along with their significance when searching data. The `sourcetype` field plays a crucial role in Splunk as it helps classify and categorize data by its source type. Splunk uses a pre-trained list of source types to automatically detect and assign the appropriate sourcetype if none is specified during the input phase. `sourcetype` definitions are configured in the `props.conf` file, where data administrators create custom ones based on the type of data they're working with. `sourcetype` configuration consists of important settings such as line breaking boundaries, timestamp extraction, and charset encoding. These settings are applied during the input, parsing, and indexing phases that we briefly examined.

Splunk Web's **Add Data** feature is a great way to test the data input by uploading a small amount of data and applying the sourcetype settings. We looked at the steps involved to upload an access log data file to our Splunk instance. We also saw how the Add Data feature offers other ways to index data through forwarders and by monitoring local files on the Splunk instance.

In the next chapter, we will deep-dive into the Splunk data inputs that we were briefly introduced to at the beginning of this chapter.

Self-assessment

The questions and options provided in this section are similar to the question pattern used in the certification exam. If you have difficulty answering any question, feel free to revisit the corresponding section in this chapter. Best of luck!

1. Choose the data indexing phases (Select all that apply).

   A. Indexing

   B. Parsing

   C. Input

   D. Ingestion

2. Which phase involves data collection and contains various data input options for indexing?

   A. Parsing

   B. Input

   C. Ingest

   D. Indexing

3. In which phase is the license calculated?

   A. Input

   B. Parsing

   C. Indexing

   D. Ingest

4. You are a data administrator working with data stored in files on a source system running the latest version of a Linux OS. Which of the following methods for Splunk data input would you choose?

   A. Write a script to monitor the file and forward it directly to the indexer

   B. Install a UF and configure file-based monitoring

   C. Transfer the file to an FTP where the indexer polls for a file to ingest

   D. Write a script to read each line and send it to the HEC

5. What is the data input type in Splunk that exposes a RESTful API to accept data for indexing?

   A. The network data input sends events over an API

   B. A HEC exposes a RESTful API for data indexing

   C. Scripted inputs integrate with Splunk RESTful Data Input APIs

   D. File monitoring can read and integrate with a RESTful API for indexing

6. Identify the metadata fields. (Select all that apply)

   A. `time`

   B. `host`

   C. `sourcetype`

   D. `source`

   E. `index`

7. Is it possible to override the metadata fields set in the input phase in subsequent phases while indexing data?

    A. Yes

    B. No

8. Once the index is set in the input phase, it cannot be overridden. True or false?

    A. False

    B. True

9. Which feature of Splunk allows you to test the sourcetype settings by uploading a data sample?

    A. Add Data

    B. Upload Data

    C. Batch Upload

    D. Input Data

10. Which official website contains add-ons and apps built by Splunk?

    A. `splunkapps.com`

    B. `splunkbase.com`

    C. `splunkappstore.com`

    D. `splunkappaddon.com`

Reviewing answers

1. *Options A, B, and C* are the correct options. There is no phase called ingestion.

2. *Option B* is the correct answer. The input phase offers the default data input types supported by Splunk, and you can additionally install add-ons to collect data.

3. *Option C* is the correct answer. License usage is calculated in the last phase of the data ingestion process, that is, the indexing phase.

4. *Option B* is the correct answer. File monitoring through a UF agent is the best approach of the options available.

5. *Option B* is the correct answer. The HEC is an agentless input type that exposes a RESTful API to accept data from external sources for indexing.

6. *Options B, C, D, and E* are the correct default fields. There is no `Time` field – instead, `_time` is the correct name.

7. *Option A, Yes*, is correct. Default fields such as `index`, `host`, `source`, and `sourcetype` can be overridden in the parsing phase.

8. *Option A, False*, is the correct answer. The index set in the input phase can be overridden in the parsing phase.

9. *Option A* is the correct answer. The **Add Data** feature in Splunk Web allows us to upload sample data of up to 500 MB in size for creation/sourcetype testing. The upload feature is also useful for one-shot ingestion.

10. *Option B* is the correct answer. `https://www.splunkbase.com` contains a huge number of apps and add-ons built by Splunk and the community developers.

9
Configuring Splunk Data Inputs

Getting data into Splunk Enterprise is the primary responsibility of a data administrator. There are multiple ways to get data into Splunk, including the standard data inputs that are popular and used across a range of data input sources. In this chapter, we will learn about these data inputs in more detail, including the suitability of these inputs with regard to data sources, and how to create monitoring inputs and adjust the configuration settings.

We'll cover the following topics in this chapter:

- File and directory monitoring
- Network inputs (TCP/UDP)
- Scripted inputs
- **HTTP Event Collector** (**HEC**) aka agentless data input
- Windows inputs

We explored these data inputs briefly in *Chapter 8, Getting Data In*. Splunk Enterprise is built for data, it works on data, and it returns data for various business use cases. Data administrators involved in getting data into Splunk must adopt the correct approach, set metadata accurately, and apply appropriate parsing settings. We will learn more about parsing in *Chapter 10, Data Parsing and Transformation*.

The majority of the data input types in the preceding list require a **Universal Forwarder** (**UF**) on the source machine. I suggest installing a UF to test the configuration settings that we will discuss in the following sections.

We have covered *Installing the universal forwarder* section in *Chapter 4, Splunk Forwarder Management* for both Windows OS and Linux OS environments. Alternatively, you can follow the steps outlined in the Splunk docs to install a Windows UF at `https://tinyurl.com/2p83uc3v`, and similarly, for operating systems such as Linux, Solaris, macOS, FreeBSD, and AIX, you can follow the steps documented at `https://tinyurl.com/f4e8hrvr`.

After installation, log in as an admin to your Splunk Enterprise instance, as installed in the *Standalone* section of *Chapter 1*. Once logged in, navigate to **Settings | Forwarding and receiving | Configure receiving | New Receiving Port**, then enter 9997 in the listening port field, and click **Save**.

Then, configure the UF to connect to the receiving indexer (in this case, Splunk Enterprise) either by editing the outputs.conf file directly or via the command line, as we saw in the *Configure forwarding* section of *Chapter 4, Splunk Forwarder Management*. For more details, you can refer Splunk documentation: https://tinyurl.com/2p8h9nuc.

The following diagram details the UF and indexer (the Splunk Enterprise instance) setup that we are trying to create.

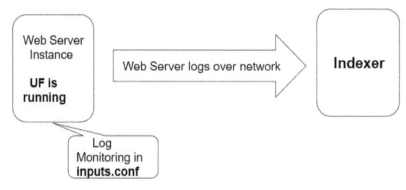

Figure 9.1: The UF sending logs to a receiving indexer

Great! You now have a UF agent ready for you to configure inputs. Let's begin with file and directory monitoring data input.

File and directory monitoring

As the name implies, this input type monitors the data files and the directories they're stored in. This is the most effective way to bring in data and is recommended by Splunk as one of the best approaches to handling files. The monitoring settings are configured in the inputs.conf file. To enable this input, a UF agent is required on the source machine, and the same settings also work on **Heavy Forwarders** (**HFs**) and Splunk Enterprise. Let's look at the notable features of this input type:

- Works for all text-based files including structured formats (XML, CSV, JSON, etc.) and .gzip-compressed files.

- Keeps track of the files being monitored via checkpoints maintained in a fishbucket directory, under $SPLUNK_HOME/var/lib/splunk/.

- Resumes the file and directory monitoring from the last location in the event of forwarder restarts.

- Recursively discovers all the files in a directory, including any new files created.

- Uncompresses the files before monitoring and forwarding them.

- Able to discover file rotation and won't monitor files that are renamed through rotation by a parent process.

- Its available configuring interfaces are the Splunk CLI, configuration files, and Splunk Web options. Splunk Web is only available on the HF and Splunk Enterprise.

- The character set in Splunk while forwarding data for indexing is vital to support internationalization. The CHARSET setting can be defined on a forwarder in a props.conf file. By default, on non-Windows-based systems, it is set to UTF-8; on Windows, it is set to AUTO. Refer to the props.conf.spec file for more details.

Let's look at the inputs.conf monitoring specification for Unix-like operating systems. The first line that starts with [monitor:// is called a *stanza*. We discussed configuration files in *Chapter 6, Splunk Configuration Files*:

```
# inputs.conf file monitor high-level specification for *nix Operating
systems,
[monitor://<path to file (or) path to directory>]
<key> = <value>
```

The full specification is available at https://tinyurl.com/5a2v5epy, or in both Splunk Enterprise and the Splunk UF installation at $SPLUNK_HOME/etc/system/README/inputs.conf.spec.

Take a look at the following Windows file monitoring configuration – we can see the monitor:// stanza followed by an absolute path pointing to the WindowsUpdate.log file. The host and sourcetype metadata fields have also been set:

```
#inputs.conf file on a windows host
[monitor://C:\Windows\System32\WindowsUpdate.log]
host = myexamplehost
sourcetype = windowsupdate
index = windows-os
disabled = 0
```

Take a look at the following file-monitoring configuration on a Linux host – the monitor:// stanza is followed by an absolute path pointing to the messages file. The host and sourcetype metadata fields have been set:

```
## inputs.conf file on a Linux host
[monitor:///var/log/messages]
host = myexamplelinuxhost
sourcetype = linux_messages
index = linux-os
disabled = 0
```

Alternatively, Splunk offers CLI commands to add functionality to monitor files and directories. Under the hood, it adds configurations to the `inputs.conf` file in the `$SPLUNK_HOME/etc/apps/search/local` directory.

The command to add `/var/log/messages` to monitoring is as follows – execute it under the `bin` directory of the Splunk/forwarder installation:

```
./splunk add monitor /var/log/messages -index linux-os
```

If you recall *Chapter 8*'s *Data indexing phases* section, in *Figure 8.2*, the initial input stage involves monitoring and processing inputs through the settings specified in the `inputs.conf` file. This critical phase is responsible for defining the data sources and inputs that will be collected and indexed. The `<key>` = `<value>` pairs in this file, also called settings, whose specification is defined in the `inputs.conf.spec` file, are vast in number. We are going to only focus on the metadata settings in this file that are heavily used during search time by Splunk users.

Particularly important metadata fields configurable during the input phase are `host`, `sourcetype`, `source`, and `index`. Let's look at the settings in `inputs.conf` that are related to these fields:

```
# inputs.conf monitoring specification to set metadata
[monitor:///file/on/hostname1/path/is/webserver-h1.log]
host = <name of host>
host_segment = <To extract host from segment in file path>
host_regex = <To extract host from the path to file name. If host_
segment and host_regex both set then host_segment takes precedence and
host_regex is ignored by input>
sourcetype = <custom sourcetype>
index = <name of index to forward logs>


#inputs.conf settings to override the host details through host_
segment
[monitor:///file/on/hostname1/path/is/webserver-h1-access.log]
host_segment = 3
sourcetype = webserver:access
index = web


#inputs.conf settings to override the host details through host_regex
[monitor:///file/on/hostname1/path/is/webserver-h1-access.log]
host_regex = \w+-(\w+)-\w+.log
sourcetype = webserver:access
index = web
```

In the previous two examples of `inputs.conf`, one had `host_segment=3` extracted and set `host=hostname1` (extracted from the file path segments separated with /), and in the other example, `host_regex` set `host=h1` (extracted from the file path that matches the `webserver-h1.log` portion). If both settings are present under one monitoring stanza, then `host_segment` takes precedence over `host_regex`.

`index` is set to `web`, while `sourcetype` is set to `webserver:access`.

There are more interesting settings useful in various scenarios that we will examine now. An up-to-date list of settings is available at `https://tinyurl.com/5a2v5epy`:

- `...` notation (three-dot notation/ellipsis) in the `monitor` path results in the recursive monitoring of subdirectories. For example, `[monitor:///file/on/hostname1/.../is/webserver.log]` traverses all the subdirectories recursively under the `hostname1` segment and looks for files named `webserver.log`.

- Using the `*` notation (wildcard) in the `monitor` path matches elements within a specific path segment, including files within that segment. Usually, this notation is used at the very end of a path or as a suffix in segments to match directories and files. For example, `[monitor:///file/on/hostname*/is/webserver-*.log]` matches on directory segments starting with `hostname` and filenames starting with `webserver-`.

- The `ignoreOlderThan` setting is used to exclude from monitoring those files older than a specified time. For example, `ignoreOlderThan = 10d` won't monitor files with last-modified dates of older than 10 days. Valid units for the time parameter are non-negative integers in the syntax of `[d|h|m|s]`, representing days, hours, minutes, and seconds respectively.

- `whitelist` and `blacklist` are filters used to specify which files or directories should be included or excluded from monitoring. In this context, the blacklist takes precedence over the whitelist, meaning that if a file or directory is listed in both the blacklist and whitelist, it will be excluded from monitoring.

- The `_TCP_ROUTING` option allows the forwarding of data to specific `tcpout` groups that are, in turn, pointed to indexers. `tcpout` groups are configured in the `outputs.conf` file to enable forwarding. See the `inputs.conf.spec` and `outputs.conf.spec` files for more details.

That's pretty much it in terms of important settings to monitor files and directories.

A distinct feature of Splunk with regard to this input is that it keeps track of files under continuous monitoring by storing seek pointer and **Cyclic Redundancy Check** (**CRC**) information in a `fishbucket` directory, which is helpful to avoid indexing duplicate data. In some scenarios, the data might need to be re-indexed; this could be from either a particular source or a reset of all the sources at once. The following list of Splunk CLI commands are useful for resetting the fishbucket for re-indexing:

- Reset a specific source:

  - Go to `$SPLUNK_HOME/bin` and execute the following command:

    ```
    ./splunk cmd btprobe -d    $SPLUNK_HOME/var/lib/splunk/
    fishbucket/splunk_private_db    --file <source> --reset
    ```

- Here's an example of resetting the `/var/log/messages` source:

```
./splunk cmd btprobe -d    $SPLUNK_HOME/var/lib/splunk/
fishbucket/splunk_private_db    --file /var/log/messages --reset
```

- Reset the whole fishbucket, which affects all sources on that host – this can be done in a couple of ways:

  - Execute the following command:

```
./splunk clean eventdata index _thefishbucket
```

  - Remove the fishbucket directory under `$SPLUNK_HOME/var/lib/splunk/`

> **Important note**
>
> Make sure the Splunk/forwarder instance is stopped while executing these commands. After successful completion of these commands, you may need to start the service.

That's a lot to take in, isn't it? I recommend checking the `inputs.conf.spec` and `inputs.conf` files under the `$SPLUNK_HOME/etc/system/default` directory to familiarize yourself with the settings. Alternatively, you could practice the preceding examples by creating an `inputs.conf` file in Splunk or by installing the UF agent. In the next section, we will look into the network data input settings.

Handling network data input

The network data input type is available for sources that can only send data over TCP/UDP. Sources such as IoT devices, network switches, routers, and sensors rely on TCP/UDP layer-4 protocols, the indexing of data from which is supported by Splunk Enterprise. Here are some important details about this input type:

- The UF and HF both support network input.

- In Splunk Enterprise, the indexer instance is usually preceded or "fronted" by the UF or HF to handle the task of forwarding data for indexing. The connection from UF/HF to Splunk Enterprise must use a valid **Socket Secure Layer** (**SSL**) certificate.

- **Transmission Control Protocol** (**TCP**) is more reliable than **SSL User Data Protocol** (**UDP**), as the latter doesn't guarantee the delivery of network packets.

- UDP messages in Splunk are not indexed as individual events until a timestamp is found in the data stream. This can be fixed during the parsing phase by configuring `sourcetype` to use the `SHOULD_LINEMERGE=false` setting in the `props.conf` file on an HF/indexer.

- Syslog messages and **Simple Network Management Protocol** (**SNMP**) traps are two types of data that need additional setup for indexing.

- As a best practice, Splunk recommends forwarding syslog messages to a central syslog collector for centralized storage. The syslog collector receives and stores these messages as files on the filesystem. To ingest this data in Splunk, you can employ the file and directory monitoring data input method that we discussed previously in this chapter. It continuously monitors the stored files and fetches new data as it is appended to those files.

- SNMP traps, which are valuable for monitoring network devices, can be ingested into Splunk using various methods. In the case of traps being stored in a file, you can employ the file and directory monitoring data input method. Check out splunkbase.com for SNMP trap collectors.

- TCP/UDP inputs provide settings for the index, sourcetype, host, and source metadata fields. Splunk recommends not overriding the source field, as the input phase will determine the appropriate value for problem analysis and investigation.

While these facts are still fresh in our minds, let's look at the network input specification for both TCP and UDP.

TCP and UDP input

The following settings go into the inputs.conf file on a UF/HF host that is dedicated to receiving network data by the system admin (the full specification is available at https://tinyurl.com/28kjyu6a):

```
## inputs.conf TCP network Input specification
[tcp://<remote server>:<port>]
connection_host = <determines host field ip|dns|none>
acceptFrom = <comma separated TCP addresses/CIDR ranges, default *
matches all and !(not) to reject connections>
index = <name of index to forward data>
sourcetype = <sourcetype name>
host = <override if connection_host=none>
queueSize = <in-memory queue size notation -
integer[KB|MB|GB],    Example value 500KB>
persistentQueueSize = <queue to store on disk when in-memory is full
notation - integer[KB|MB|GB|TB], Example value 100MB >
```

In the preceding code, the [tcp://<remote server>:<port>] stanza listens on the TCP network port and accepts traffic from the specified remote server. If no remote server is specified, [tcp://<port>] accepts data from any remote host. Splunk suggests using the acceptFrom setting to allow/restrict specific remote hosts instead of remote server as a best practice.

acceptFrom is used to set allow or deny rules for remote hosts. By default, it is set to *, meaning it accepts connections from everywhere. To reject a connection, use ! ("not" notation) followed by the remote host name pattern.

In the event of network issues or the indexer being unavailable, we can prevent data loss by setting `queueSize` and `persistentQueueSize`. The `queueSize` setting refers to the in-memory queue that holds the data until the output queue on the forwarder is free. The `persistentQueueSize` setting enables the storage of data on disk when the in-memory queue is full. `persistentQueueSize` must be greater than `queueSize`. Data is written to the `$SPLUNK_HOME/var/run/splunk/` directory on the forwarder.

The UDP input specification is the same as for TCP, except that it listens on the UDP port – the rest of the settings and syntax remain the same. An example `inputs.conf` configuration for UDP is as follows:

```
# inputs.conf listening on UDP port 514
[udp://514]
connection_host = ip
acceptFrom = 192.0.2.10, !external.example.host.com
index = main
sourcetype = access_common
queueSize = 1MB
persistentQueueSize = 100MB
```

Network inputs might have restrictions on *nix and Windows-based hosts, as they require access to certain ports to listen on TCP/UDP. Try it yourself by configuring these inputs on your UF/Splunk Enterprise instance and checking the ports' listening status (use the `netstat` command on Linux and Windows-based hosts).

We have reached the end of the network inputs section. In the next section, we will go through the configuration of scripted inputs.

Discussing scripted inputs

Scripted inputs are useful for indexing transient/temporary data that cannot be monitored through file/directory monitoring or use network inputs. Scripted inputs collect data from transient sources and then either write the collected data to a file or forward it directly to an indexer. Let's go through some facts about this input type:

- Scripted inputs require a UF agent or Splunk Enterprise instance (HF) to execute the scripts.

- Data can be gathered from transient sources, such as operating system commands. The `top`, `vmstat`, `netstat`, and `iostat` commands all leverage this type of input, which is configured within the Splunk add-on for Unix and Linux available to download from `http://splunkbase.splunk.com`. The Windows **technology add-on** (**TA**) relies on this input type to gather Windows **Active Directory** (**AD**) logs, registry logs, `WinEventLogs`, and so on. Logs from remote APIs can also be pulled using scripts.

- Popular script types are supported, including `.py` (Python), `.sh` (shell), `.ps1` (PowerShell), and `.bat` (Windows batch).

- Scripted inputs must be scheduled to run at certain intervals. How often a script should run is set by specifying the interval as a number of seconds or by setting a `cron` expression.

- Splunk Enterprise and the UF log script execute messages in a local `splunkd.log` file. You can get access to these logs in the `_internal` index.

- Scripted inputs are configurable through Splunk Web under the **Settings | Data Inputs | Scripts** menu. Another option is to create them in the `inputs.conf` file, optionally package them into an app, and use a deployment server to deploy them to forwarders. Refer to *Chapter 4, Splunk Forwarder Management.*

Consider the following scripted input specification:

```
# inputs.conf scripted input settings
[script://<script path or cmd>]
source = <source to set>
sourcetype = <sourcetype to set>
index = <index name to forward logs>
interval = <in seconds. default 60 sec or cron expression. Set -1 to
run once, 0 to continuously>
passAuth = <username to run the script as the OS user>
queueSize = <in-memory queue size>
persistentQueueSize = <queue to store on disk when in-memory is full>
```

The preceding specification has the stanza name prefixed with `script://`, followed by the usual common metadata fields of `source`, `sourcetype`, and `index`, and the queue settings. The settings specific to scripted input are `interval` and the authentication information.

The `interval` setting schedules the script to run every *x* number of seconds. If nothing is specified here, the script will be scheduled to run every minute by default. Alternatively, we could specify a cron expression.

`queueSize` and `persistentQueueSize` work similarly, as explained in the *Handling network data input* section – in the event of indexer unavailability, both settings hold the data generated by the script in memory and on disk.

The `passAuth` setting ensures the script will be run by the assigned username by creating an authentication token for that user. Before setting up a scripted input in Splunk, you need to create the script and ensure it is located in a filesystem path accessible to Splunk. The `[script://<script path>]` stanza will then be able to execute the script from that designated location. For security reasons, scripts must be placed in one of the following locations:

- `$SPLNUK_HOME/etc/apps/<app name>/bin`

- `$SPLNUK_HOME/etc/system/bin`

- `$SPLNUK_HOME/bin/scripts`

However, there is an exception for scripts with the `.path` extension, which in turn can refer to scripts (`.py`, `.sh`, `.ps1`, and `.bat`) anywhere on the host. The `.path` script file itself must be in one of the preceding locations, but the scripts it refers to can be located outside of these directories.

Let's look at the following scripted input that invokes the *nix `top` command to get the process information. Here the script path is relative to the `inputs.conf` file. The script has the `.sh` extension and is scheduled to run every 60 seconds. `sourcetype` and `source` are both set to `top`. The logs generated by this script will be forwarded to a custom index named os (standing for *operating system*):

```
[script://./bin/top.sh]
interval = 60
sourcetype = top
source = top
index = os
disabled = 0
```

If you want to refer to more such examples, download the `Splunk_TA_nix` add-on from `https://splunkbase.splunk.com/app/833` and examine the `/default/inputs.conf` and `/bin` directories.

> **Tip**
> The best practice is to place scripts close to the `inputs.conf` file. For example, if the file is in the `$SPLNUK_HOME/etc/system/local/` directory, then place the script in the `$SPLNUK_HOME/etc/system/bin/` directory. Likewise, if `inputs.conf` is in the `$SPLNUK_HOME/etc/apps/<app name>/local` directory, then place the script in the `$SPLNUK_HOME/etc/apps/<app name>/bin` directory.

That is all on scripted inputs for now – try installing an add-on from Splunkbase that uses scripted inputs to test this functionality. In the next section, we will look at HEC input.

Understanding HEC input

HEC is an agentless input type that doesn't require a forwarder on the source machine. This input type is suitable for sources that are capable of sending events over HTTP(S), such as web apps (through JavaScript libraries), mobile apps, and automation scripts. HEC exposes RESTful API endpoints on the Splunk Enterprise instance to accept data for indexing. The instance could be an HF/indexer in a distributed deployment. Let's look at the key facts about HEC input:

- HEC is disabled by default on Splunk instances, and the user must enable it manually to start using it. HEC can be scaled by configuring it across multiple Splunk instances and optionally fronting it with a load balancer.

- Authentication to HEC APIs is done via a token supplied in the HTTP request sent by the source. The token configuration is set up by the administrator and shared with the application/source team for them to send over the events.

- HEC exposes two important RESTful endpoints for data indexing – one to access JSON-only formatted events and the other to accept any raw type of data. The JSON RESTful endpoint is designed to accept data that is formatted only in JSON. When sending data to this endpoint, you should ensure that the payload (data) is properly formatted in the JSON structure with additional metadata fields as required. The original data event that is wrapped inside the JSON payload can be a raw event

- Each token created for every application/source must be a unique identifier. The best practice is to use 32-character **Globally Unique Identifiers (GUIDs)** as tokens, such as CD1E753F-5BB2-4F5C-AE22-FE0381A0D916. The source application presents this token in HTTP requests while sending events as a payload to the collector endpoint.

- HEC auto-parses the JSON events received, as their event timestamps and other metadata follow a predefined structure. For raw data, the line-breaking strategy is applied if it contains a timestamp; alternatively, the sourcetype settings can be configured to override this behavior accordingly in props.conf.

- HEC offers an event-indexing confirmation feature through the useAck indexer acknowledgment setting. Indexer acknowledgment is enabled for individual tokens at the stanza level; a separate endpoint provides the status of the event indexing.

- To allow/restrict certain addresses from sending data over HEC, the acceptFrom setting can be configured at the [http] stanza level in the inputs.conf file.

Configuring HEC

We will now go through the step-by-step process to configure HEC.

Step 1: Let's prepare the ground to start sending data over HEC. The first step is to enable the HEC in Splunk Web. Navigate to **Settings | Data Inputs | HTTP Event Collector** and click on the **Global Settings** button in the top-right corner. This will open a form with the **All tokens** option. Set this option to **Enabled** and click **Save**.

You should have noticed a couple of things while enabling HEC in **Global Settings**, one of which is that **Enable SSL** is checked by default, and the other being that the HTTP port number where collector endpoints listen to receive the events is 8088.

HEC global settings are also configurable via the [http] stanza, located by default at $SPLUNK_HOME/etc/apps/splunk_httpinput/default/inputs.conf. To enable HEC via this method, add the following setting to the $SPLUNK_HOME/etc/apps/splunk_httpinput/

`local/inputs.conf` file and restart the Splunk instance to reflect the change. We do this because Splunk advises the creation of a `/local` directory to change the settings:

```
[http]
disabled = 0
```

Step 2: We are ready to create an HEC token. A token can be created in three different ways – through Splunk Web, the Splunk CLI, or by directly editing the `inputs.conf` file. We will create the token by editing the `.conf` file as follows, using the example of a fictional application team named `mobile-5g`:

```
# inputs.conf HEC token stanza settings
[http://token-for-mobile-5g]
description = To receive events from 5G networks
disabled = 0
host = StandaloneSplunkInstance
index = mobile_5g
token = 09776ade-cf23-42c0-9138-89ad8388516a
sourcetype = signling_data
useACK = 1
```

In the preceding code, we can see a `[http://<name>]` stanza with the `index`, `sourcetype`, and `host` metadata fields, which can be overridden by the application client sending events to HEC along with a token in GUID format. The `useACK` setting is set to 1, meaning that indexer acknowledgment is enabled for this token. If the application team wants to find the status of indexing, they can do so. The token will be shared with the fictional `mobile-5g` application team for them to send signaling data.

To create a token through Splunk Web, go to **Settings** | **Data Inputs** | **HTTP Event Collector** and click on the **New Token** button in the top-right corner, which will open a page with three stages to be completed, starting with **Select Source**. The required inputs for each stage clearly correlate with the preceding `inputs.conf` settings. Completing those stages will create a token.

Sending data to HEC

We now have a token created for the `mobile-5g` application. Applications in general use the HTTP client libraries, available through programming languages such as Java, .NET, JavaScript, Python, and so on, to format payloads and send them to the HEC API endpoints.

The following are the HEC API endpoints that the majority of application clients use:

- `https://<splunk-hostname>:8088/services/collector/raw`: Receives any type of text/raw event.
- `https://<splunk-hostname>:8088/services/collector/event`: Receives JSON formatted only event.

- `https://<splunk-hostname>:8088/services/collector/ack`: This is used to query the event indexing status. It works for tokens that have indexer acknowledgment enabled.

- The full list is available at `https://tinyurl.com/j53xmynj`.

To showcase how to send data to these HEC endpoints, we will use the `curl` command in a Linux environment. We can see there are two header values – one with `Authorization` carrying the token value, and the other with `\X-Splunk-Request-Channel` having a unique GUID specific to the client.

Here we are sending data as a JSON-formatted payload – a collector/event request:

```
curl -k -H "Authorization: Splunk 09776ade-cf23-42c0-9138-
89ad8388516a" -H "X-Splunk-Request-Channel: FE0ECFAD-13D5-401B-847D-
77833BD77131" https://mysplunk.example.com:8088/services/collector/
event -d '{"sourcetype": "signaling_data", "event": "stable signal!"}'
```

The following is the collector/event response:

```
{"text":"Success","code":0,"ackId":0}
```

Here we are sending data as a raw event – a collector/raw request:

```
curl -k -H "Authorization: Splunk 09776ade-cf23-42c0-9138-
89ad8388516a" -H "X-Splunk-Request-Channel: FE0ECFAD-13D5-401B-847D-
77833BD77131" https://mysplunk.example.com:8088/services/collector/raw
-d 'stable signal!'
```

We get the collector/raw response as follows:

```
{"text":"Success","code":0,"ackId":1}
```

Since we have set `useACK = 1 (true)` at the token level, the client must send a unique channel identifier in the request to HEC. In return, the client receives `ackID` in the JSON format from HEC (refer to the preceding response payloads in the collector/raw event example). The client further uses these acks to query HEC's `/ack` endpoint to find the event indexing status as follows.

Finding the event indexing status – `/ack` endpoint request – requires the channel identifier and `acks` returned in response to the client in the preceding two JSON and raw event requests:

```
curl -H "Authorization: Splunk 09776ade-cf23-42c0-9138-89ad8388516a"
-H "X-Splunk-Request-Channel: FE0ECFAD-13D5-401B-847D-77833BD77131"
https://mysplunk.example.com:8088/services/collector/ack -d
'{"acks":[0,1]} '
```

The JSON response payload for the preceding `ack` endpoint request looks as follows:

`{"acks": {"0": true, "1": false}}` — here, `true` represents successful indexing and `false` is used for a failed indexing status of that event. Once `acks` has returned `true` for a particular event, its indexing status will vanish from memory, and if a client sends another request afterward, it will return `false`.

We have come to the end of HEC input – it's so technical, isn't it? Don't be overwhelmed by the technological jargon. You need to remember to treat HEC like a normal API that requires authentication by presenting a token with every request. Indexer acknowledgment is essentially the same as a parcel delivery tracking feature – for example, after you place an online order (say, through the `/raw` or `/events` endpoints), you get a tracking number in return (`ackID`), and by entering the tracking number (`acks`) into a portal (the `/ack` endpoint), you can track where your parcel is. Finally, after delivery, you get a *successful delivery* status from the tracking portal.

In the next section, we will go through the Windows inputs.

Exploring Windows inputs

The input types that we have gone through so far are neither technology- nor OS-specific. However, here, as the name suggests, Windows inputs work on Windows-only hosts. The host requires at least a UF or Splunk Enterprise instance to collect Windows-specific logs. Windows natively stores its logs in binary format, and the Windows inputs interact with OS APIs to get these logs.

When installing a UF on Windows hosts, you are given the option to enable Windows inputs. With a Splunk Enterprise instance, Windows inputs can be configured through Splunk Web via **Settings | Data Inputs** and then choosing **Local event log collection** and **Remote event log collection**. Remote event log collection requires a Windows domain account for log collection from remote hosts.

In a large-scale Windows host environment with UF already configured, use a deployment server to centrally manage inputs. Create and configure `inputs.conf` within the app. Deploy the app to the forwarders for efficient and centralized input configuration. Let's look at some key facts about Windows inputs:

- Windows inputs are able to collect Windows events, AD, performance monitoring, registry, printer, host, and network logs.

- **Windows Management Instrumentation** (**WMI**) is used to collect event logs from remote hosts in the network, typically on Windows servers with domain accounts.

- Remote event log collection is configured in the `wmi.conf` file. The full specification is available at `https://tinyurl.com/37mdekpe`.

- Enabling Windows logs can lead to significant data size increases, potentially consuming a substantial portion of our Enterprise license. Splunk provides two settings, `whitelist` and `blacklist`, to filter unwanted event logs before they even get indexed. This helps to reduce the license usage and index only required events.

- Collected log events can be forwarded to any indexer type (including one running on *nix) in a distributed Splunk deployment.

- Global settings for Windows event logs can be defined under a new [WinEventLog] stanza along with the [default] stanza. According to inputs.conf.spec, it is recommended to place global settings under both stanzas. Refer to https://tinyurl.com/bdfpsc4y for more info.

The full Windows input specification for inputs.conf.spec is available at https://tinyurl.com/yf7kzrmb. Let's look at some Windows input stanza examples for various types of logs:

```
# Monitoring Security log channel
[WinEventLog://Security]
<key> = <value>
whitelist = <comma-separated list of event codes> | key=regex
[key=regex]
blacklist = <comma-separated list of event codes> | key=regex
[key=regex]
```

An example Event Logs application log channel with event filtering is as follows:

```
[WinEventLog://Application]
whitelist = 4,5,6,7
blacklist = EventCode=%^200$%
```

The specification for performance monitoring of a local physical disk, using the perfmon:// stanza prefix, is as follows:

```
[perfmon://LocalPhysicalDisk]
<key> = <value>
```

The AD monitoring specification using the admon:// stanza prefix is as follows:

```
[admon://<name>]
<key> = <value>
```

To monitor print-related logs, the specification uses the WinPrintMon:// stanza prefix:

```
[WinPrintMon://<name>]
type = < predefined values Printer;Job;Driver;Port>
```

wmi.conf contains a stanza in this format – [WMI://<name>].

The following example illustrates the configuration settings for local memory monitoring and remote USB changes:

```
## Local memory monitoring
[WMI://localmemorymon]
wql = <wql query>
internal = <collection frequency in seconds>

## remote server USB changes
[WMI://remote-usb-changes]
wql = select * from __InstanceOperationEvent within 1 where
TargetInstance ISA 'Win32_PnPEntity' and TargetInstance.
Description='USB Mass Storage Device'
internal = 10
server = example.remotehost.com
```

As there are numerous Windows input monitoring channels, it can be difficult to remember all of them. Instead, concentrate on understanding the types of input logs that can be collected, such as event logs and AD logs. The WMI interface is beneficial for gathering data both locally and remotely. Additionally, it is important to learn how to filter events using the `whitelist` and `blacklist` settings.

We have now come to the end of this chapter. We have gone through a lot of configuration stanzas and settings for various input types. After the following summary are some practice exam-style questions covering the data input types that we have learned so far.

Summary

In this chapter, we learned heaps about configuration settings for various data inputs that are very useful when getting data into Splunk. We began with the steps to install a UF and jumped into the file and directory monitoring input, understanding how this is used to monitor files and directories recursively. We also learned about ... (the three-dot notation/ellipsis used to traverse the directories in the filesystem path recursively) and * (the wildcard notation) in a monitor file path. We understood the use of fishbucket to keep track of files monitored using checksums and how it can be reset using the `btprobe` command.

We looked into network inputs used to accept data over TCP/UDP. TCP is more reliable than UDP, although their configuration specifications are very similar. Afterward, we covered the scripted input type that executes scheduled scripts and indexes transient data. Scripted inputs are commonly utilized in numerous Splunkbase apps, necessitating adherence to specific standards and script wrapping for proper script placement.

Supported script formats for scripted inputs include `.sh` (shell script), `.py` (Python script), `.bat` (batch script), and `.ps1` (PowerShell script).

Finally, we examined the HEC and Windows input types. HEC exposes RESTful APIs for JSON formatted and raw events, and to get the status of indexing. We understood that HEC is disabled by default and worked through a step-by-step approach, starting from enabling HEC to sending data over the API. We concluded with Windows inputs, designed for Windows-only hosts to collect various log events, both locally and remotely.

Do note that out of all the inputs, HEC doesn't require a forwarder on the source machine, for which reason it is also called *agentless input*. HEC can be enabled on Splunk Enterprise instances such as an HF or as an indexer in a distributed deployment to receive log events from application clients. File and directory monitoring, scripted inputs, and Windows inputs need at least a UF on the source host. Network inputs can be enabled on UFs and Splunk instances (typically, an HF).

This chapter dealt with the input collection phase, which is the first phase as outlined in *Chapter 8, Getting Data In*. In the upcoming chapter, we will delve into the parsing phase, which involves processing the data collected through these data inputs.

It is now time to test your knowledge by answering the following questions on the data input topics that we covered in this chapter.

Self-assessment

1. Select the popular data input types offered by Splunk by default. (Choose all that apply):

   A. Binary file monitoring

   B. File and directory monitoring

   C. Network data input

   D. Scripted input

   E. Network port monitoring

2. You are about to configure a file monitoring input and observe that the directory contains five-year-old data that does not need to be indexed. Which setting is used to force the forwarder to ignore the old files?

   A. `skipHistoricalFiles`

   B. `ignoreOldData`

   C. `ignoreOlderThan`

   D. `deletePastData`

3. A network device can send traffic over UDP on port 514. You are an admin and need to allow incoming traffic from the network device IP. What setting is appropriate for this situation?

    A. `acceptOnly`

    B. `acceptFrom`

    C. `connection_host`

    D. `acceptIPOnly`

4. You have been given the task of configuring the monitoring of files with the names `network-sys-messages.log`, `sys-messages.log`, and `syslogs.log` in a directory path of `/opt/var/log/syslog/` on a Linux system. They all fit into a single source type – syslog. What is the optimal approach to configuring the appropriate monitoring on a forwarder?

    A. Create three different `monitor://` stanzas pointing to individual files.

    B. Create one monitor stanza with a wildcard * `[monitor:///opt/var/log/syslog/*sys*.log]`.

    C. Create one monitor stanza with an ellipsis `[monitor:///opt/var/log/syslog/sys....log]`.

    D. Both options A and B are optimal approaches.

5. Is HEC enabled by default on Splunk instances?

    A. Yes

    B. No

6. What is the default HEC port?

    A. 8089

    B. 8000

    C. 8088

    D. 8090

7. Which statement is true about the `useAck` setting for HEC token configuration?

    A. By setting `useAck` = `true` or `1`, the indexer acknowledgment feature is enabled.

    B. By setting `useAck` = `false`, the indexer acknowledgment feature is enabled.

    C. By setting `useAck` = `true` or `1`, the indexer will write the contents to disk.

    D. By setting `useAck` = `false`, the indexer will acknowledge the indexing status to the HEC client.

8. You need to collect Windows security event logs from multiple Windows hosts across an organization. After a few days of Windows input being enabled, it is observed that the ingestion rate is very high. The security team has advised that certain event codes do not need to be ingested. Which Windows input setting can be used to reduce the rate of ingestion in this scenario?

   A. `whitelist` the required event codes.

   B. `blacklist` the unwanted event codes.

   C. Stop the ingestion for a few days and then reenable it.

   D. Both options A and B are correct approaches.

9. Which of the following statements about fishbucket are true? (Select all that apply):

   A. It is one of the buckets associated with the `_internal` index.

   B. It keeps track of monitoring files.

   C. fishbucket auto-resets after every restart of the forwarder.

   D. fishbucket is a directory.

10. You have been given the task of ingesting logs from a source that doesn't support any of the data inputs offered by Splunk by default. Where can you go to find other mechanisms for data collection?

    A. Splunk doesn't support other types of collection mechanisms.

    B. `splunkbase.com` might have an add-on related to the source's technology.

    C. A Splunk support case is required to find such mechanisms.

    D. Splunk suggests not ingesting data from sources other than the default data inputs, so you should fit your data into one of them.

Hopefully, you are able to recollect the topics covered. Let's review the answers as follows.

Reviewing answers

1. *Options B, C, and D are correct. Options A and E* don't exist in Splunk.

2. *Option C is correct.* Historical files are ignored when monitoring by setting `ignoreOlderThan` in the `inputs.conf` file of the forwarder.

3. *Option B is correct.* `acceptFrom` is the recommended setting to restrict clients' sending of data over network ports.

4. *Option B is correct.* `*` (wildcard) matches potentially an unlimited number of characters. The three-dot/ellipsis (…) notation is used to work with directories recursively.

5. *No*. HEC isn't enabled by default. The admin needs to enable and configure it for use.

6. *Option C* is correct. `8088` is the default HEC port.

7. *Option A* is correct. `UseAck` = `true` or `1` enables indexer acknowledgment, which tracks the indexing status of messages sent over HEC. A separate endpoint (ending with `/ack`) is used to query the indexing status of an event by supplying the `ackID` value returned in response to the HEC request.

8. *Option D* is correct. Both blacklist and whitelist can be used.

9. *Options B and D* are correct. `fishbucket` is a directory that keeps the tracking information of files monitored.

10. *Option B* is correct. If the default data input types are of no use in a given scenario, then `splunkbase.com` is the best place to look for an appropriate add-on. TAs are available for log collection from a variety of technology sources, such as Linux, AWS, Azure, Google cloud, and ServiceNow.

10
Data Parsing and Transformation

The first phases of the data journey is the input phase, which we discussed in detail in *Chapter 9, Configuring Splunk Data Inputs*. Data parsing is the second phase, followed by data being indexed on the disk. This chapter deals with the parsing phase, which comes right after the input phase and ends by handing over the data to the index phase for storage and preparation for data searching.

The question that might arise is what the need for the parsing phase is, as all the data has been collected, the metadata fields are set during the input phase, and finally, data is forwarded to indexers for indexing. The prominent features of the parsing phase are breaking the whole data stream into individual events, extracting and applying timestamps, setting the metadata fields to individual events, manipulating metadata before indexing, and transforming the data if needed. During the input phase, metadata fields such as the index, host, sourcetype, source, and charset are applied to the entire data and not to individual events. The data stream is processed as a whole, and no line-breaking or timestamp extraction is done during the input phase.

Events are transformed in the parsing phase. Example use cases could be overriding the sourcetype, host metadata, and data routing to a different index, extracting index-time fields, and masking or anonymizing sensitive information such as **Personally Identifiable Information** (**PII**), bank accounts, **Social Security Numbers** (**SSNs**), and phone numbers.

In this chapter, we are going to learn about parsing and transformation settings by applying them through the `props.conf` and `transforms.conf` files. As a Splunk data administrator, knowing about this phase is crucial to applying the right solution to the use cases that might arise when onboarding data, or at a later point in time. For example, you might be asked to mask credit card information, override the source type, change the event-breaking boundaries, or re-route specific events to a different index altogether. We are going to learn about the following topics in this chapter:

- Understanding the parsing phase
- Parsing phase settings – `props.conf`

- Transformation settings – `transforms.conf`
- Splunk Web data preview

Let's get started.

Parsing phase settings

Going through the parsing phase is crucial before data is indexed. Parsing happens right after the input phase – the process is input -> parsing -> indexing, in this order. You can refer to the *Data indexing phases* section of *Chapter 8*, which introduced these phases, for a refresher. The data stream must be preprocessed before indexing. The parsing phase in Splunk is necessary for formatting and extracting relevant data from unstructured or semi-structured input, making the data searchable and actionable. The following are some of the important sub-phases that data goes through:

- Breaking the whole data stream into individual events
- Identifying the timestamp of an event if needed and applying it
- Applying metadata fields such as the host, sourcetype, source, and index
- Optionally transforming the data by re-routing, overriding metadata, masking portions of events, filtering and dropping unnecessary events, and so on

These sub-phases are configured through two files in Splunk, which are the `props.conf` and `transforms.conf` files. The settings related to these sub-phases will be discussed in the next sections. In a distributed deployment, either a **Heavy Forwarder** (**HF**) or an indexer works with these parsing settings. **Universal Forwarders** (**UFs**) do not have the ability to parse data; however, structured files (such as CSV, PSV, JSON, and XML) are the exception to this. If the data traffic to indexers is routed through an HF, it is always advisable to deploy these settings to the HF itself.

The single/standalone Splunk instance that we installed in the *Splunk installation – standalone* section of *Chapter 1* will be used here for testing. Let's begin with the `props.conf` file settings. Before we begin, let's read about when to restart/reload after making changes to these conf files.

From now on, the changes that we are going to make to the configuration files require a reload/restart of Splunk Enterprise/the forwarder in the following scenarios. Since we have installed a standalone Splunk instance, reloading Splunk Enterprise is sufficient. To reload, hit this URL – `http(s)://<yoursplunkserverhost>:8000/en-GB/debug/refresh` – and enter your admin credentials. Deployments with HFs need to be restarted after changes to conf files have been made. If the **deployment server** (**DS**) is used for the deployment of apps to forwarders, then the DS can be configured to auto-restart HFs after a successful deployment. However, it is less likely that restarting/reloading will be required for Splunk Enterprise when changes are made via Splunk Web, a REST API, or the Splunk CLI. When to restart the search head/indexer is covered in the Splunk docs; refer to `https://tinyurl.com/2e9mke2e`.

props.conf settings

If you want to refer to the full **specification** (**spec**) for the props.conf file, it is available here – https://tinyurl.com/epznpja3. You won't be tested on all the settings that you find in the specifications file; just go through the settings that we are going to discuss in this chapter.

Some of the parsing phase settings require **regular expressions** (**regex**); I highly recommend you read up on the basics of these. There are a number of online resources available for reference. To test a regex, I prefer to use https://regex101.com/, which is free to use. Splunk's regex engine uses the **Perl Compatible Regular Expression** (**PCRE**) library, written in C under the hood. When using the https://regex101.com/ website, you also use PCRE.

The props.conf settings are applied to the data stream by identifying it by its metadata fields. The metadata fields are specified in stanzas in the following format:

- **The sourcetype field**: The settings defined under this sourcetype stanza apply to all data streams regardless of their index, host, or source. Splunk comes with default source types to reuse, such as access_combined, apache_error, and cisco_syslog. Some data sources do not fit into the existing source types; hence, you must create custom ones. Using the following stanza, we will look at creating a new web_access source type:

  ```
  [<sourcetype-name>]
  <key> = <value>
  ```

  As an example, in props.conf, consider the TRUNCATE parsing phase setting, which is designed to restrict the maximum line length in bytes that Splunk processes.

  For example, by setting TRUNCATE to 1000 for the access_common source type, Splunk restricts any lines in the data that match the specified sourcetype to a maximum line length of 1000, in bytes. That means the remaining lines exceeding 1,000 bytes will be truncated:

  ```
  [access_common]
  TRUNCATE = 1000
  ```

- **The source field**: The [source::<source>] stanza is used to define a specific source or source-matching pattern for an event. By specifying a source or pattern, Splunk can apply custom configurations and formatting rules to data from that source during the parsing phase:

  ```
  [source::< regex style source-name pattern>]
  <key> = <value>
  ```

As another example, when you set TRUNCATE to 1000 for a specific source, such as /var/log/web-server/access.log, this means that any lines in the data that match that source pattern will be truncated to a maximum length of 1,000 bytes. Setting TRUNCATE to 1000 for source restricts the maximum line length in bytes for sources that match the specified pattern:

```
[source::/var/log/web-server/access.log]
TRUNCATE = 1000
```

- **The host field**: The [host::<host or host pattern>] stanza is useful for defining the custom configuration settings for data originating from the given host or host pattern. These settings could be related to data formatting, extraction rules, and many other things according to the props.conf specification during the parsing phase:

```
[host::< host or host pattern >]
<key> = <value>
```

For example, when you set TRUNCATE to 1000 for a specific hostname, such as myhost.com, this means that any lines in the data that originate from the myhost.com host, will be truncated to a maximum length of 1,000 bytes:

```
[host::myhost.com]
TRUNCATE = 1000
```

- **Stanza precedence**: When multiple stanzas match a given event, the settings defined in the most specific stanza take precedence over those defined in less specific stanzas. Specifically, settings defined in [host::<host or host pattern>] stanzas override those in [<sourcetype name>] stanzas, and settings defined in [source::<source or source pattern>] stanzas take priority over both [host::<host or host pattern>] and [<sourcetype name>] stanzas. This hierarchy of settings allows fine-grained control over how data is processed in Splunk, ensuring that the most relevant settings are applied to each event.

At this stage, we know about stanzas that are useful in various scenarios. Let's have a look at the parsing settings specification for line breaking, timestamp identification, line merging, and so on. I am going to use the sourcetype stanza to describe these settings. For example, our sourcetype name is web_access. Do note that settings key names are case-sensitive:

```
## Parsing settings for web_access sourcetype
[web_access]
LINE_BREAKER = < regex: By default, set to number of new line or
carriage returns ([\r\n]+) >
SHOULD_LINEMERGE = <true | false - To merge several multi-line events
into one event. If set to true it depends on the number of other
settings. Read full spec>
BREAK_ONLY_BEFORE = <Applicable only if SHOULD_LINEMERGE set to true>
BREAK_ONLY_BEFORE_DATE = <Applicable only if SHOULD_LINEMERGE set to
```

```
true>
TIME_PREFIX = < Regex: If set, Splunk software scans the event
text for a first match of this regex before attempting to extract a
timestamp.>
TIME_FORMAT = <strptime-style format>
MAX_TIMESTAMP_LOOKAHEAD = < Integer: The number of characters into an
event Splunk software should look
for a timestamp. By default, looks for first 128 characters >
TZ = <Timezone of the event, UTC, AEDT, AST and so on. If not set
splunkd process timezone is considered by default.>
TRANSFORMS-<class name> = <transforms.conf stanza1>, <transforms.conf
stanza2>
```

While the preceding settings can be useful for optimizing Splunk's data processing, it's worth noting that Splunk will attempt to use default values whenever possible. These defaults are designed to provide a reasonable level of performance and functionality for most use cases. The TRANSFORMS-<class name> setting is relevant to transforms.conf, which we will discuss in the next section. Let's categorize these settings by their function:

- **Line breaking**: The LINE_BREAKER setting breaks the data stream into individual lines of events. By default, it looks for the number of new lines or carriage returns to find the event boundary. SHOULD_LINEMERGE must be set to false if LINE_BREAKER is set, and this is advisable for performance reasons.

- **Line merging**: SHOULD_LINEMERGE combines multiple events into a single event. You can set this to true or false; by default, it is set to true. When we set this to true, other settings such as BREAK_ONLY_BEFORE_DATE and BREAK_ONLY_BEFORE must be set to identify the event boundary.

- **Timestamp identification**: It is highly important to configure the timestamps that are applied to every event in the parsing phase correctly. Splunk has a set of predefined date-time formats ($SPLUNK_HOME/etc/datetime.xml), which are popular and standard. By default, if an explicit TIME_FORMAT is set, Splunk scans the first 128 characters to find the timestamp in an event. If the event contains a timestamp beyond 128 characters, or to restrict timestamp scanning from the TIME_PREFIX position in an event, you could use the MAX_TIMESTAMP_LOOKAHEAD or TIME_PREFIX setting. TIME_FORMAT is a strptime standard style format that can be set if the timestamp is in a non-standard format. Refer to the date-time format variables here – https://tinyurl.com/3kaddz23. TZ identifies the time zone the data originated from. To view a list of all the time zone (TZ) IDs, see https://en.wikipedia.org/wiki/List_of_tz_database_time_zones.

There are scenarios in which the event might not have a date, time, or year value, or may have none of these. Splunk tries to determine these by following its own rules to identify timestamps in exceptional cases. The Splunk documentation for *How timestamp assignment works* is available here – https://tinyurl.com/4tjaaerb.

Let's look at the following example of web_access logs to define the props.conf parsing settings. The logs have timestamp information; -0500 in every log line represents the **Eastern Standard Time** (**EST**) time zone and every new line in the logs represents the event boundary. Let's write the props. conf information for this web_access log sample:

```
127.0.0.1 - - [10/Feb/2023:10:34:56 -0500] "GET /example-page.
html HTTP/1.1" 200 2345 "https://www.example.com/" "Mozilla/5.0
(Windows NT 10.0; Win64; x64) AppleWebKit/537.36 (KHTML, like Gecko)
Chrome/96.0.4664.110 Safari/537.36"
127.0.0.1 - - [10/Feb/2023:10:37:56 -0500] "GET /example-page.
html HTTP/1.1" 400 2345 "https://www.example.com/" "Mozilla/5.0
(Windows NT 10.0; Win64; x64) AppleWebKit/537.36 (KHTML, like Gecko)
Chrome/96.0.4664.110 Safari/537.36"
127.0.0.1 - - [10/Feb/2023:10:39:56 -0500] "GET /example-page.
html HTTP/1.1" 500 2345 "https://www.example.com/" "Mozilla/5.0
(Windows NT 10.0; Win64; x64) AppleWebKit/537.36 (KHTML, like Gecko)
Chrome/96.0.4664.110 Safari/537.36"
```

The following props.conf configuration breaks the access logs into individual events by every new line (the default behavior) and applies the correct timestamp with a time zone. The main goal is to parse the access logs correctly:

```
#props.conf for preceding web_access logs
[web_access]
LINE_BREAKER = ([\r\n]+)
SHOULD_LINEMERGE = false
TIME_PREFIX = \[
TIME_FORMAT = %d/%b/%Y:%H:%M:%S %z
TZ = EST
MAX_TIMESTAMP_LOOKAHEAD = 26
```

The [web_access] stanza provided is an example of the configuration in the props.conf file in Splunk. Here's what each setting does:

- LINE_BREAKER = ([\r\n]+): This setting specifies the regex pattern used to identify line breaks in the data. In this case, it is set to ([\r\n]+), which matches one or more occurrences of either a carriage return or a line feed character.

- SHOULD_LINEMERGE = false: This setting controls whether or not Splunk should attempt to merge consecutive lines of data into a single event. By setting this to false, you are telling Splunk to treat each line of data as a separate event.

- TIME_PREFIX = \[: This setting specifies the regex pattern used to identify the beginning of a timestamp in the data. In this case, it is set to \[, which matches a literal opening square bracket character.

- `TIME_FORMAT = %d/%b/%Y:%H:%M:%S %z`: This setting specifies the format of the timestamp in the data. The `%d/%b/%Y:%H:%M:%S %z` format string corresponds to a timestamp in the format of `day/month/year:hour:minute:second timezone`.

- `TZ = EST`: This setting specifies the time zone to use when parsing the timestamp. In this case, it is set to EST.

- `MAX_TIMESTAMP_LOOKAHEAD`: This setting works in conjunction with `TIME_PREFIX`. After the `TIME_PREFIX` regex match, the timestamp extraction process starts. `MAX_TIMESTAMP_LOOKAHEAD` specifies the maximum number of characters from the position immediately following the `TIME_PREFIX` match, where the timestamp actually begins, that Splunk will consider when attempting to extract the timestamp. In this case, the `TIME_PREFIX` match ends with the character [right before the timestamp starts (for example: `10/Feb/2023:10:34:56 -0500`). The `MAX_TIMESTAMP_LOOKAHEAD` value of 26 allows Splunk to look 26 characters beyond the position where the `TIME_FORMAT` match ends to find the complete timestamp, including the time zone information.

Together, these settings define how Splunk should process data from a source that follows the `web_access` log format. Out of all these settings, we have one thing left, which is `TRANSFORMS-<class name>`. In the next section, we are going to learn about it, along with the other settings in the `transforms.conf` file. Let's get into it.

Transformation settings – transforms.conf

In this section, we are going to learn about the `transforms.conf` file settings and their relationship to the `props.conf` file. The link between these two files is a reference to the setting in `TRANSFORMS-<class name>` in the `props.conf` file. Defining the stanzas in `transforms.conf` alone is of no use without referring to them in the `props.conf` file. The full specification for `transforms.conf` is available here: `https://tinyurl.com/24cbp5u4`.

You might be wondering about its purpose. There are many tasks that we cannot do alone with the `props.conf` file during the parsing phase. The `transforms` settings are used during parsing as well in the following scenarios:

- `transforms.conf` stanzas can be reused by referring to them in many other sources, source types, hosts, and so on across the same Splunk platform. Define them once and use them many times by referring to them in the `props.conf` file.

- Data anonymization is possible through the `transforms.conf` file. SEDCMD is the setting used inside `props.conf` to anonymize the data without the need for the `transforms.conf` stanza.

- Configure host, source, and source type overrides.

- Filter and route specific events to a particular index.

- Filter and drop unwanted events using `nullQueue`.

- Create new index-time field extractions.

- Create advanced search-time field extractions.

- Set up lookup tables that look up fields from external sources.

In the next section, we are going to learn about data manipulation use cases, such as data anonymization, re-routing data to a different index, filtering events, and overriding a source type. We will learn about the remaining field extractions and lookups in *Chapter 11, Field Extractions and Lookups*. Let's begin with data anonymization.

Data anonymization

Data anonymization is the process of transforming PII into a format that cannot be used to identify the individuals to whom the information is originally linked. The purpose of data anonymization is to protect the privacy and confidentiality of sensitive data.

There are several methods for data anonymization, and one of them is masking, which we are going to learn about in this section.

Masking involves replacing sensitive data with non-sensitive placeholder values – for example, replacing a person's phone number with XXXX or their date of birth with a randomized value. Do note that once the data is masked during the parsing and index phases, that cannot be undone.

Splunk has the ability to mask data using the following two approaches. As I mentioned before, it is highly recommended to have prior knowledge of regex to understand this section:

- The SEDCMD setting in props.conf alone

- Using both props.conf and transforms.conf

Using SEDCMD in props.conf

Applying the SEDCMD setting through props.conf directly works on raw individual events. You could use this setting for the sourcetype, source, and host stanzas as follows. If you are aware of sed in Linux, it works in a similar way to the sed command in Linux. The specification of it is as follows:

```
# props.conf SEDCMD setting specification
[<sourcetype-name> | source::<source name> | host::<hostname>]
SEDCMD-<class> = <sed like command>
```

The <sed like command> syntax is explained as follows:

- sed like command – s/regex/replacement/flags.

- <class> is a unique literal string.

- s – for substitution.

- The regex in sed is a PERL-compatible regex style (optionally containing capturing groups).

- The replacement is a string to replace the regex match. Use \n for back references, where n is a single digit.

- Flags can be either g to replace all matches or a number to replace a specific match.

To be clearer, let's go through the following example of call log data that includes phone numbers:

- **Example**: Company A is a telecom company indexing the call record data to Splunk Enterprise with index set to carrier-records and sourcetype set to call_records. The call_records data contains PII such as phone numbers, which must be masked for privacy reasons. As data administrators, our task is to use the SEDCMD setting we just learned about to mask the phone numbers with XXXXXXXX symbols at the sourcetype level. We could make use of the standalone Splunk instance that we installed in the *Splunk installation – standalone* section of *Chapter 1* for this example.

- **Log sample**: In the following call records sample, you shall be able to find phone numbers in the format (555) 555-1234, which we are going to mask using SEDCMD as follows.

Here's the call_records log sample:

```
2022-03-13 15:22:08 Srikanth (555) 555-1234 made a call to Jane (555)
555-5678
2022-03-12 10:11:24 Mary (555) 555-4321 sent a message to Scott (555)
555-8765
2022-03-11 18:05:03 David (555) 555-2345 received a call from Abdul
(555) 555-9876
```

The props.conf file's SEDCMD setting for masking is as follows:

```
## props.conf masking phone numbers in call records
[call_records]
SEDCMD-maskphonenumber = s/\(\d{3}\)\s\d{3}-\d{4} /XXXXXXXX /g
```

In the props.conf file provided, SEDCMD specifies a regex pattern for phone numbers and the replacement string to be used in their place. The g flag indicates that all occurrences of phone numbers in the logs should be replaced with the specified replacement string (XXXXXXXX).

Once this configuration is deployed to the standalone Splunk instance, any logs that are ingested and match the specified regex will be transformed accordingly and indexed in their transformed state. However, it's worth noting that logs that were previously indexed prior to this configuration change will not be affected by this transformation.

If you need to mask phone numbers in previously indexed logs, you'll need to re-ingest those logs after the configuration change has been deployed.

Here are the `call_records` logs after masking:

```
2022-03-13 16:22:18 Chuppa XXXXXXXX made a call to Chand XXXXXXXX
2022-03-13 11:19:54 Garry XXXXXXXX sent a message to Bill XXXXXXXX
2022-03-13 22:55:23 Dan XXXXXXXX received a call from Ahmad XXXXXXXX
```

You can see that the phone numbers have been masked with XXXXXXXX by issuing a search request in Splunk – `index=carrier-records sourcetype=call_records`.

In the next section, we are going to learn about data masking using `transforms.conf` and `props.conf` settings.

Using transforms and props

We are going to use the same `call_records` log sample here and apply masking settings using the `transforms.conf` and `props.conf` files. Let's look at the specification:

```
# transforms.conf specification
[<transforms-stanza-name>]
SOURCE_KEY = <_raw by default, _MetaData:Index, MetaData:Sourcetype,
MetaData:Host, MetaData:Source and so on>
REGEX = <regular expression to operate on field contents specified in
SOURCE_KEY>
FORMAT = <specifies the format of the event works in accordance with
REGEX, for fields format is fieldname::fieldvalue>
DEST_KEY = <_raw for masking in data event, can set other keys as
described in Spec>

## props.conf specification
[<sourcetype-name> | source::<source name> | host::<hostname>]
TRANSFORMS-<class> = <transforms stanza name 1>, <transforms stanza
name 2>.. <transforms stanza name N>
```

The `transforms.conf` spec looks pretty simple; however, it supports several use cases that are out of scope for this book. The number of key combinations that can be used to manipulate metadata fields such as `host`, `sourcetype`, `source`, and `queue` are all described in the spec file here – `https://tinyurl.com/27sxf94z`.

The `props.conf` spec has the `TRANSFORMS-<class>` setting. Here, `<class>` is a unique literal string. We can specify more than one `transforms` stanza and comma-separate them to apply them in the same order.

We are going to use the preceding `transforms.conf` and `props.conf` specifications for the rest of the use cases described in the following sections, which deal with overriding the source type, re-routing data to a different index, filtering unwanted events, and sending them to `nullQueue`.

Let's focus on our original goal of masking the phone numbers in the `call_records` log using `transforms.conf`. The settings are as follows:

```
## transforms.conf stanza for masking
[mask-phonenumber-in-call-records]
SOURCE_KEY = _raw
REGEX =     ^([\d\-:\w\s]+)\(\d+\)\s+\d+-\d+.+?([\d\-:\w\s]+)\(\d+\)\
s+\d+-\d+.+?$
FORMAT = $1 ********* $2 *********
DEST_KEY = _raw

## props.conf referring to transforms stanza for sourcetype
[call_records]
TRANSFORMS-masknumber = mask-phonenumber-in-call-records
```

Once these configurations are deployed to our standalone Splunk instance, any logs that are ingested as a `call_records` source type and match the specified regex will be transformed accordingly and indexed in their transformed state. However, it's worth noting that logs that were previously indexed prior to this configuration change will not be affected by this transformation. By looking at the preceding `transforms.conf` settings, a regex is used to match phone numbers in the `_raw` dataset. The matched phone numbers are then replaced with asterisks (`*`), and the resulting string is written to DEST_KEY as specified.

If you need to mask phone numbers in previously indexed logs, you'll need to re-ingest those logs after the configuration change has been deployed.

Here are the `call_records` logs after masking:

```
2022-03-13 16:22:18 Chuppa ********* made a call to Chand *********
2022-03-13 11:19:54 Garry ********* sent a message to Bill *********
2022-03-13 22:55:23 Dan ********* received a call from Ahmad *********
```

You can see that the phone numbers in the newly ingested data have been masked with asterisks (`*********`). That is it for data anonymization – we masked sensitive information in the `call_records` logs using both approaches, with `props.conf`, SEDCMD, and `transforms.conf`.

In the next section, we will go through overriding source types on a per-event basis.

Overriding source types

`sourcetype` (remember, Splunk fields are case-sensitive) is a default metadata field name. If you recall from the earlier section on the `props.conf` settings, the `sourcetype` definition is applied by Splunk to format the data and apply timestamps to individual events during the parsing phase.

The best practice is to set `sourcetype` during the input phase if possible. It could be set on the UF, in the **HTTP Event Collector** (**HEC**) payload, and in other input types wherever possible. The following are a few situations in which overriding the source type might be required:

- Splunk's automatic source type detection might not have selected the right type

- The default out-of-the-box `sourcetype` settings might not parse the data stream correctly

- There could be more than one file with the same name from different sources but often the contents differ

Overriding the source type during the parsing phase is achievable via the `transforms.conf` settings. The specification that we learned about in the preceding section using SOURCE_KEY, DEST_KEY, REGEX, and FORMAT involves the main settings that are required.

The scenario that we are assuming here for writing the transforms is that data is originating from a source system running a UF that is monitoring input – `/var/log/syslog/messages` is set to the `linux_secure` source type, which is not correctly parsing the data stream. Hence, the requirement is to set the correct `sourcetype` that has been already defined – `syslog_messages`.

Here's what we know about the data stream:

```
source - /var/log/syslog/messages
```

The default `sourcetype` detected by Splunk is `linux_secure`.

The correct `sourcetype` to override it with is `syslog_messages`.

Let's look at both transforms; the props configurations are as follows:

```
## transforms.conf stanza for sourcetype override
[override-with-syslog-messages]
SOURCE_KEY = _raw
REGEX = .
FORMAT = sourcetype::syslog_messages
DEST_KEY = MetaData:Sourcetype
## props.conf referring to transforms stanza
[source::/var/log/syslog/messages]
TRANSFORMS-set_sourcetype = override-with-syslog-messages
```

In the preceding example, we defined a new transformation called `override-with-syslog-messages` that will match all events (REGEX = .), and FORMAT (syntax `<fieldname::fieldvalue>`) is working in accordance with REGEX and applying `syslog_messages` to the `sourcetype` field as required. After the successful deployment of these settings, every new event from that moment is set to the new source type, `syslog_messages`, originating from `/var/log/syslog/messages`. Already ingested events won't be changed by this action and `sourcetype` changes are only applied to new events from when the transforms settings were deployed.

If you want to give it a try, you could override the `host` field with the required hostname value. In that case, `DEST_KEY` would be `MetaData:Host`. In the next section, we are going to learn about index routing.

Index re-routing

An index is a repository of data being set during the input phase. There might be situations in which some of the events in the data stream should be filtered and re-routed through a different index. This is achievable in the parsing phase and works only on an HF/indexer.

Some applicable situations could be as follows:

- The data stream contains security data that is required to be filtered and stored in a separate index. Storing it in a separate index gives more flexibility in terms of access control and retention.

- If the data size is large enough, a single index might not work efficiently for search processing. In that case, routing a portion of data that's being accessed frequently to a high-performance index and retaining less frequently used data in a low-performance index would improve the search performance.

- Due to compliance requirements, some events should be stored in a separate index.

To achieve index routing, we are going to rely on `transforms.conf`. Its specification remains the same as we have seen so far in previous sections.

The example scenario I am considering here is the `call_records` log sample that we used in the *Data anonymization* section. If you look carefully at the log sample, it contains calls made between customers, as well as **Short Messaging Service (SMS)** records.

The requirement is to separate the SMS records and route them to `index: sms-conversations`. The *message* and *sent* keywords in the logs are going to be the REGEX match used to identify SMS records:

```
## Log sample
2022-03-13 15:22:08 Srikanth (555) 555-1234 made a call to Jane (555)
555-5678
2022-03-12 10:11:24 Mary (555) 555-4321 sent a message to Scott (555)
555-8765
2022-03-11 18:05:03 David (555) 555-2345 received a call from Abdul
(555) 555-9876
```

Let's look at the `transforms.conf` settings, as follows:

```
## transforms.conf stanza for index routing
[sms-index-routing]
SOURCE_KEY = _raw
REGEX = (message|sent)
FORMAT = sms-conversations
```

```
DEST_KEY = _MetaData:Index

## props.conf referring to transforms stanza
[call_records]
TRANSFORMS-set_sms_index = sms-index-routing
```

In the given example, the regex (REGEX) is applied to the _raw field, which includes the message or sent keyword. Based on the defined FORMAT and DEST_KEY, any matching events are directed to the sms-conversation index. After the successful deployment of these settings, the call_records events related to SMS are indexed as part of the sms-conversations index.

In the next section, we are going to learn about filtering and dropping unwanted events.

Dropping unwanted events

Organizations index massive amounts of data within Splunk to get answers out of it. Often, admins need to cut down the amount of unwanted data getting indexed. The reason for this is huge amounts of data require and consume the license and extra resources for execution, and after all, if the data is not going to be used for any business use case, then there is no point in indexing it in Splunk.

transforms.conf provides an opportunity to filter such unwanted data and drop it before it is indexed. This all happens during the parsing phase and only works on an HF/indexer.

To set the ground, I am considering the same call_records log sample. The requirement is to drop the frequently occurring phone number call record that ends with 5678:

```
## call_records log sample
2022-03-13 15:22:08 Srikanth (555) 555-1234 made a call to Jane (555)
555-5678
2022-03-12 10:11:24 Mary (555) 555-4321 sent a message to Scott (555)
555-8765
2022-03-11 18:05:03 David (555) 555-2345 received a call from Abdul
(555) 555-9876
```

Let's look at the transforms.conf settings:

```
## transforms.conf stanza to drop events match
[drop-high-profile-phonenumbers]
SOURCE_KEY = _raw
REGEX = (\(\d{3}\)\s\d{3}-5678)
FORMAT = nullQueue
DEST_KEY = queue

## props.conf referring to transforms stanza
[call_records]
TRANSFORMS-drop_5678 = drop-high-profile-phonenumbers
```

In the preceding example, the regex (REGEX) is applied to the _raw field, which is matched to events that include a phone number ending with 5678. Based on the defined FORMAT and DEST_KEY, any matching events are directed to nullQueue, which discards/drops the whole matching event. After the successful deployment of these settings to the HF, the following two call_records events are indexed and the event matching 5678 is dropped:

```
2022-03-12 10:11:24 Mary (555) 555-4321 sent a message to Scott (555)
555-8765
2022-03-11 18:05:03 David (555) 555-2345 received a call from Abdul
(555) 555-9876
```

So far, we have learned about the transforms.conf and props.conf settings by configuring them via our desired text editor. In the next section, we are going to learn how to test the source type settings in real time using the Splunk data preview option through the UI.

Splunk Web data preview

The Splunk Web data preview, **Add Data**, and **Upload** features are useful for testing the source type settings defined according to the props.conf specification. This works for the [<source type>] stanza.

Why is this so important? For example, say you create a new source type with line-breaking and timestamp extraction settings through a text editor or the **New Source Type** option available on Splunk Web at **Settings | Source types** and deploy it. Afterward, you realize the data indexed is not correctly formatted. You have to troubleshoot and find the invalid setting and then redo the same deployment again and again until everything is right. It is tedious, isn't it? The data preview feature comes in handy for testing the source type and transforms.conf settings. The transforms. conf settings cannot be directly tested in the UI; however, we could pre-create them and refer to them when testing the source type settings. Remember that transforms.conf stanzas are always referred to in the props.conf file.

What do you need for data preview?

- A working Splunk Enterprise instance to get hands-on with. We learned about installation in *Chapter 1* in the *Splunk installation – standalone* section.

- A sample data file that you want to index.

- Knowledge of regex.

- Awareness of the props.conf and transforms.conf settings discussed in previous sections.

Let's get into it – I am taking the example of the call-records data sample once again for testing. We are going to test the source type definition and masking using SEDCMD in this section.

The `call-records.sample.log` file I have prepared for testing has the following contents in it:

```
2022-03-13 15:22:08 Srikanth (555) 555-1234 made a call to Jane (555)
555-5678
2022-03-12 10:11:24 Mary (555) 555-4321 sent a message to Scott (555)
555-8765
2022-03-11 18:05:03 David (555) 555-2345 received a call from Abdul
(555) 555-9876
2022-03-13 15:22:08 Srikanth (555) 555-12345 made a call to Geenie
(555) 555-56789
2022-03-12 10:11:24 Mario (555) 555-54321 sent a message to Scoot
(555) 555-87656
2022-03-11 18:05:03 Sannro (555) 555-23456 received a call from Jeff
(555) 555-98765
```

In the next section, we are going to create a new source type definition for the `call_records` sample data and test it using the Splunk Web data preview feature.

Creating the source type definition

In this section, we will upload the `call-records` log sample to test and apply the source type settings. Let's go through the steps:

1. Log in to Splunk Web and navigate to **Settings**, then click on **Add Data**:

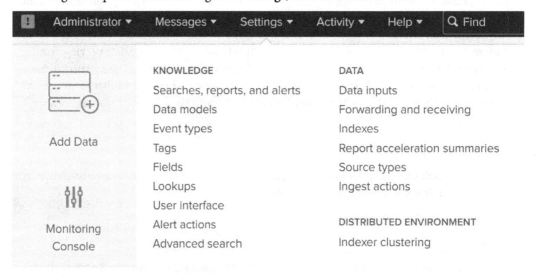

Figure 10.1: Splunk Web | Settings | Add Data

Figure 10.1 shows the **Add Data** option to select. Clicking on it will take you to the following screen, as shown in *Figure 10.2*.

2. Scroll to the bottom of the page and you will find the **Upload** option:

Figure 10.2: The Upload option for getting data in

Figure 10.2 shows three different options for getting the data into Splunk. Choose the first option, **Upload**, to upload the `call-records` sample data. Clicking will take you to the screen shown in *Figure 10.3*.

3. As you can see in *Figure 10.3*, you can use the **Select File** button or you can drop in the `call-records` sample data. I clicked on **Select File** and then opened File Explorer in Windows to locate the `call-records.sample.log` file. Click on the **Next >** button:

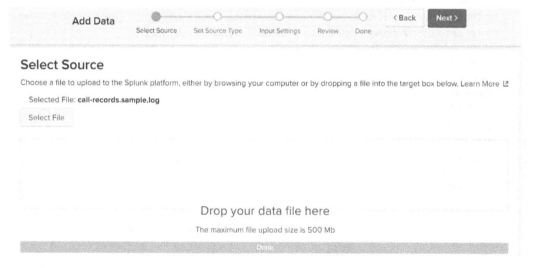

Figure 10.3: Add Data | Select File to upload the file

Figure 10.3 shows the selected file: `call-records.sample.log`. The maximum file size that can be uploaded is 500 MB.

4. Clicking on the **Next >** button will take you to the **Set Source Type** screen, as shown in *Figure 10.4*.

Set Source Type

This page lets you see how the Splunk platform sees your data before indexing. If the events look correct and have the right timestamps, click "Next" to proceed. If not, use the options below to define proper event breaks and timestamps. If you cannot find an appropriate source type for your data, create a new one by clicking "Save As".

Source: **call-records.sample.log**

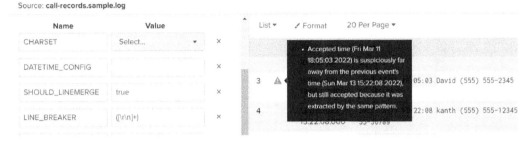

Figure 10.4: Set Source Type – default settings and data preview

Figure 10.4 shows the advanced source type settings on the left side, chosen by Splunk by default, and on the right side, the `call-records` log sample timestamp shows a warning. We are going to fix the **Name** and **Value** settings on the left side to get rid of the timestamp warning.

5. See *Figure 10.5* showing the updated source type **Name** and **Value** settings. I updated the `TIME_PREFIX` and `TIME_FORMAT` settings to fix the timestamp warning and updated `SHOULD_LINEMERGE = false`. Click on **Apply settings** to see how they affect the data on the right-hand side. As you can see in *Figure 10.5*, the timestamp warning is gone and the events are perfectly broken down into individual events on new lines.

Set Source Type

This page lets you see how the Splunk platform sees your data before indexing. If the events look correct and have the right tir "Next" to proceed. If not, use the options below to define proper event breaks and timestamps. If you cannot find an appropriat your data, create a new one by clicking "Save As".

Source: **call-records.sample.log**

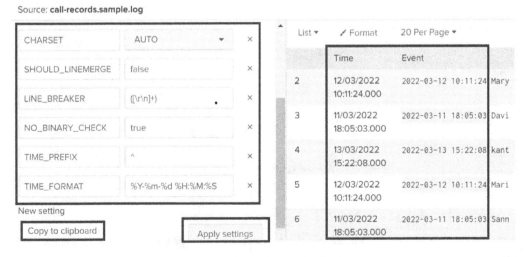

Figure 10.5: Set Source Type – Apply settings

Clicking on **Copy to clipboard** in *Figure 10.5* opens a hyperlink and displays equal `props.conf` file settings to the ones shown next. After satisfactory testing, the same settings usually get deployed to the HF/indexer in a distributed deployment. Be sure to change the `<source type>` stanza name from `__auto_learned__` to a suitable name when deploying it. In this case, as we are working with a call records sample, I would set the source type to `call_records`.

Here are the `props.conf` settings generated by Splunk Web:

```
[ __auto__learned__ ]
SHOULD_LINEMERGE=false
LINE_BREAKER=([\r\n]+)
NO_BINARY_CHECK=true
CHARSET=AUTO
TIME_PREFIX=^
TIME_FORMAT=%Y-%m-%d %H:%M:%S
```

In the case of a standalone/single instance, the settings remain deployed to the same instance as it contains the full parsing pipeline and is able to receive and index data. To create a custom source type permanently through Splunk Web, navigate to **Settings | Source Types** and click on **New Source Type**.

In the upcoming section, we will be testing the `SEDCMD` setting for data masking using the data preview feature in Splunk Web.

Data masking

In this section, we will learn about applying `SEDCMD` to mask customer phone numbers in the `call_records` data through the Splunk data preview feature and see it in action. This exercise is the same as that from the previous section, *Using SEDCMD in props.conf*.

In this section, we are going to use the Splunk data preview feature to test the masking configuration, whereas in the previous section, we directly set up the `SEDCMD` setting by editing the `props.conf` file.

Follow *steps 1-5* as described in the previous section, *Creating the source type definition*. As shown in *Figure 10.6*, the `SEDCMD` setting was added by clicking on the **New setting** hyperlink. The regex value assigned to `SEDCMD-maskphone` will match the phone number in the event. By clicking on **Apply settings**, the phone numbers in the events on the right-hand side will be replaced with XXXXXXXXX. That shows us the regex that we have written following the `sed` command style is working as expected.

Clicking on **Copy to clipboard** shows the `props.conf` settings that are created by the data preview from the **Name** and **Value** pairs configured in *Figure 10.6*:

```
# props.conf by clicking on 'Copy to clipboard'
[ __auto__learned__ ]
SHOULD_LINEMERGE=false
LINE_BREAKER=([\r\n]+)
```

```
NO_BINARY_CHECK=true
CHARSET=AUTO
TIME_PREFIX=^
TIME_FORMAT=%Y-%m-%d %H:%M:%S
SEDCMD-maskphone    = s/\(\d{3}\)\s\d{3}-\d{4}/XXXXXXXX/g
```

We get the following screen:

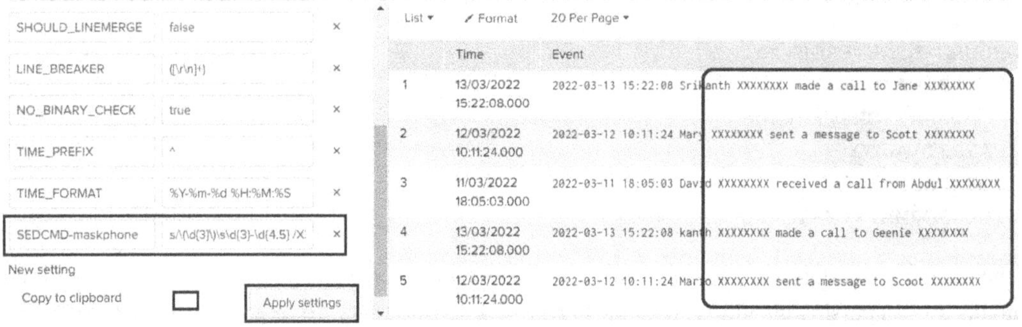

Figure 10.6: Set Source Type – SEDCMD setting

Summary

This whole chapter revolved around the `props.conf` and `transforms.conf` settings up until the end and was purely technical. We began by understanding the parsing phase, coming right after the input phase, and its significance. Out of the three components, the full parsing pipeline exists on the HF and the indexer, and not on the UF; however, the UF is able to parse structured files through the `INDEXED_EXTRACTIONS` setting. We learned that it is mandatory to deploy parsing settings on the HF if the indexers are fronted by it.

Afterward, we looked at the `props.conf` stanzas related to `sourcetype`, `source`, and `host` and went through the specification of the source type definition for line breaker, line merging, and timestamp identification. To continue, we learned about the `transforms.conf` stanzas that work in accordance with `props.conf`. We further advanced into applying the SEDCMD setting via `props.conf` for data masking and the `transforms.conf` settings for overriding the source type, index routing, and filtering unwanted events using `nullQueue`.

Finally, using the Splunk data preview feature, we tested the new source type definition settings and the data masking setting, SEDCMD.

In the upcoming chapter, we will delve into the topic of fields and explore how to create both search-time and index-time fields. Additionally, we will focus on lookups and their creation using Splunk Web and the Splunk App for Lookup File Editing.

Self-assessment

In this section, there are 10 questions followed by answers for you to test your learning throughout this chapter. If you run into difficulty in answering any of the questions, do refer to the respective section:

1. What is the next phase in data indexing right after the input phase?

    A. Indexing

    B. Parsing

    C. Masking

    D. Routing

2. Identify the stanzas that can be created in `props.conf`. (Select all that apply.)

    A. `[source]`

    B. `[sourcetype_name]`

    C. `[source::<source_name_or_pattern>]`

    D. `[host]`

    E. `[host::<hostname_or_pattern>`

3. What are the parsing phase capabilities? (Select all that apply.)

    A. Override a source type

    B. Re-route to a different index

    C. Data masking

    D. Drop events that do not need to be indexed

4. What is the name of the queue to be set for dropping unwanted events from indexing?

    A. `sinkHole`

    B. `deadQueue`

    C. `nullQueue`

    D. `dropQueue`

5. The `SEDCMD` setting belongs to the `transforms.conf` file. Is this statement true or false?

    A. True

    B. False

6. What is the limit of the file upload size in Splunk Web for the **Set Source Type**/data preview feature?

    A. 5 GB limit

    B. 1 GB limit

    C. 500 MB limit

    D. 1,500 MB limit

7. Is the source type creation saved to the `props.conf` file?

    A. Yes

    B. No

8. What setting is used to refer to the `transforms` stanza in the `props.conf` file?

    A. `TRANSFORMS-<class_name>`

    B. `TRANSPOSE-<class_name>`

    C. `TRANSFORMATION-<class_name>`

    D. `PROPS-<class+name>`

9. Do the masking settings deployed today apply to previously indexed data?

    A. Yes

    B. No

10. What setting during the parsing phase breaks the data stream into individual events with new lines as line terminators?

    A. `LINE_BREAKER`

    B. `TRUNCATE`

    C. `BREAK_EVENTS`

    D. `BREAK_ONLY_NEW_LINE`

Reviewing answers

I hope you recall the concepts that we learned about in this chapter through these questions. Let's review the answers:

1. *Option B* is the correct answer. The three main phases in the data indexing process are input, parsing, and indexing.

2. *Options B, C, and E* are the right answers. You could apply the `props.conf` settings through `[sourcetype_name]`, `[source::<source_name_or_pattern>]`, or `[host::<hostname_or_pattern>]` stanza syntax.

3. *All options* are correct. During the parsing phase, metadata fields such as `sourcetype`, `host`, and `source` can be overridden, and the index can be altered for re-routing specific events matching certain criteria and data masking. In addition, you can filter events that meet certain criteria and then drop those events from indexing.

4. *Option C* is correct. The remaining options are incorrect; no such queues exist in Splunk.

5. The answer is false. The `SEDCMD-<class>` = `"<sed style regular expression>"` setting belongs to the `props.conf` file.

6. *Option C* is correct. 500 MB is the file size limit for uploads via the **Set Source Type**/data preview feature in Splunk Web.

7. *Yes*, that's right. The source type definition will be saved to the `props.conf` file.

8. *Option A* is correct. In the `props.file` setting, `TRANSFORMS-<class_name>` = `<transform_stanza_1>,< transform_stanza_2>,...<transform_stanza_n>` is the right syntax. `<class_name>` is a unique literal string. More than one `transforms` stanza can be assigned; if you separate them with a comma (`,`), they will be executed in the specified order.

9. *No.* Data masking is a parsing phase action, and it only applies to newly ingested data in Splunk. Already ingested data from the past will not be affected. To apply the same parsing phase settings to historical data, it must be deleted and re-ingested.

10. *Option A* is correct. `LINE_BREAKER` allows you to specify how the raw data stream should be broken into individual events. By default, Splunk breaks the raw data stream into events based on the new line terminator (`\n`).

11

Field Extractions and Lookups

Great! You have reached the last chapter in *Part 2*, *Data Administration*. So far in *Part 2*, we have learned about getting the data into Splunk, adding input, and the parsing phase settings, and we have understood the phases of data traversal before it is written to disk. What we haven't seen so far is the search phase, which is fundamental for all the work we have done so far in system admin (setting up Splunk) and right after, in the data admin part (tidying up the data and storing it in indexers).

After all, if you have everything set up right and data is indexed correctly, the users who are going to search the data are going to be the real business outcome. For example, if we have indexed sales, API logs, and system logs into Splunk, the respective users could use data to generate a monthly sales report and send it via email, and alert when an API isn't available for service and/or security teams have found a vulnerability of a certain process in a system in the network. The use cases are endless for various stakeholders.

All of that work is done on a **search head** (**SH**), which is a Splunk component that issues search requests to indexers and consolidates results by exposing the **user interface** (**UI**) (Splunk Web) for reports, alerts, dashboards, and other useful features. In this chapter, we are going to learn about the fundamental building blocks known as **fields** in Splunk and how to extract them from raw data at index time and search time, and also how to enrich the data in Splunk with additional information using **comma-separated values** (**CSV**) and **Key-Value Store** (**KV Store**) **lookups**. If you are (or are going to be) a data administrator, you will work with fields almost every day.

In this chapter, you will learn about the following topics:

- Understanding fields and lookups
- Creating search-time field extractions
- Creating index-time field extractions
- Creating lookups

Let's begin with understanding fields and lookups.

Understanding fields and lookups

In this section, we will look at fields and lookups in detail. Let's start with fields.

Fields

Fields in Splunk tell a story about data that can be used to search and derive the required outcomes, such as reports, alerts, and dashboards. Raw data in Splunk is indexed as individual events by its source type definition. Fields are names given to specific portions of data by data administrators by extracting them out of the raw data during the search-time and index-time processes. Splunk, by default, assigns predefined fields to a data source, such as host, source, sourcetype, splunkserver, _time, and so on.

For example, take call record logs, which contain phone numbers from exchanged calls at a particular point in time. Let's name the fields in the log data time_of_call, caller, callee, and duration.

You could build a report such as the number of calls per day and longest duration, and alert on numbers that are calling out to international numbers without the necessary activation of an international service. All you must do is write a search query and use the respective Splunk feature, such as reports, alerts, dashboards, and so on.

Let's look at the sample call record logs:

```
2023-04-08T09:30:15, +615551234567, +615552345678, 120
2023-04-08T10:15:42, +615551234567, +615553456789, 60
2023-04-08T11:20:30, +615552345678, +15551234567, 180
2023-04-08T12:45:10, +615553456789, +615552345678, 240
```

For the first event, if we had configured search-time field extraction, the following fields/values would have appeared in the search results:

- time_of_call = 2023-04-08T09:30:15
- caller = +615551234567
- callee = +615552345678
- duration = 120

Let's go through some must-know facts about fields:

- Field names are case-sensitive, For example, `time_of_call` and `TIME_OF_CALL` are different from each other.

- The mode of search chosen by the user affects the behavior of field extraction and fields returned in search results.

  - **Fast Mode**: In this mode, field discovery is disabled, and only default and specific fields which you search for are returned.

  - **Verbose Mode**: This mode discovers and returns all available fields.

  - **Smart Mode**: This default mode discovers and returns all fields. It behaves like fast mode for transforming commands (such as `stats`, `chart`, and so on) and verbose mode otherwise.

- Field values being used while searching are case-insensitive – for example, the `index=web_logs http_method="get"` and `index=web_logs http_method="GET"` searches produce the same results.

- Fields can have different data types of values, such as alphanumeric (a) and numbers (#). Refer to *Figure 11.1* for notation

- Fields can be extracted and stored during the indexing process. That means they get stored in index files permanently on disk, which consumes additional storage space.

- The `_raw` (raw data event), `_time` (time the event occurred), and `_indextime` (time the event was indexed) Splunk internal field names are used for internal processing by Splunk, and they are accessible to users while searching data.

- The predefined/default `host`, `index`, `linecount`, `punct`, `source`, `sourcetype`, `splunk_server`, and `timestamp` fields provide basic information about data. Each of these fields is, by default, assigned to every event.

- Splunk, by default, auto extracts **key-value** (**KV**) pairs and JSON data formats at search time on the fly. The `KV_MODE` and `AUTO_KV_JSON` settings in `props.conf` enable it to work.

- Splunk allows you to create custom field extraction during the search-time and index-time processes. The search-time field extraction process provides a UI to create regular expressions for unstructured data and structured data through delimiters. We will discuss these two types of extractions in upcoming sections.

Let's look at how fields look in the Splunk search results in the following figure:

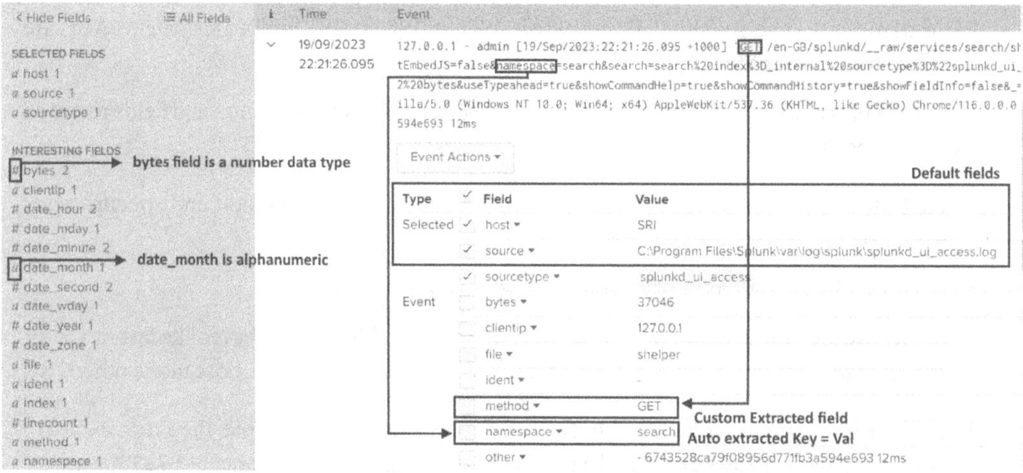

Figure 11.1: Fields in Splunk search results

The preceding figure shows how the fields appear in Splunk, which is showing default fields, a custom extraction field (`method = GET`), an auto-extracted field (`namespace = search`), and a data type symbol (`a` for alphanumeric and `#` for numeric).

Lookups

Lookups are loaded into Splunk from external sources to enrich the data with additional information. The most used lookups are structured contents in the format of CSV files and the KV Store (processed by MongoDB). Popular Splunk **Search Processing Language** (**SPL**) commands to read the lookup files are `inputlookup` and `lookup`. Splunk supports the following types of lookups. We are going to learn about CSV and KV Store-backed lookups in this chapter:

- Geospatial lookups
- External lookups
- CSV lookups
- KV Store lookups

Geospatial lookups contain geospatial information such as the longitude and latitude related to a place in the world. By default, Splunk ships with US information and worldwide countries' information. You can issue a search request in the *Search and Reporting* app – | `inputlookup geo_countries` – and click on the **Visualization** tab, and then choose **Choropleth Map**.

External lookups, also called **scripted lookups**, fetch additional information for events in Splunk by connecting to external sources such as APIs.

CSV lookups are the most popular lookups and are typically in a `.csv` format. Data admins/users define CSV files containing additional information that is required while searching data in Splunk for enrichment. CSV lookups are static in nature; hence, the best practice is to keep them simple and smaller in size. We will learn about the creation of CSV lookups in the last section.

For example, a call record log contains a country code prefixed with a number, such as +61, +1, and so on; however, it doesn't have a country name. A CSV lookup name – `Country_code_to_name_mapping.csv` – can be defined as follows and loaded into Splunk:

```
country_code, country_name
+61, Australia
+1, USA
```

A report could make use of this lookup by matching it against the country code and retrieving the country name for enhanced readability to respective stakeholders.

KV Store lookups are processed by MongoDB, which is a NoSQL document-oriented database. KV Store lookups are backed by a configuration called a **collection**. The information is stored in KV pairs in a collection and each row in a collection is uniquely identified by a field `_key`. KV Store lookups are preferred for large lookups and require frequent updates.

For example, a threat intelligence feed related to an IP address, DNS, and domain requires continuous updates to the lookup named `threat_intel`. The `threat_intel` lookup is used by security teams while triaging the potential vulnerabilities in an organization's network logs stored in Splunk.

In this section, we learned about the importance of Splunk fields for searching and retrieving data. We covered default and internal fields, as well as demonstrated the custom fields using an example of call records logs. We also discussed case sensitivity, automatic extraction of KV pairs, and Splunk Web's regular expression and delimiter extraction methods. Additionally, we explored different lookup file types, with upcoming sections focusing on CSV and KV Store lookups.

In the next section, we will learn about search-time field extractions.

Creating search-time field extractions

The SH plays a crucial role in performing search-time extractions. When a user issues a search request, the SH executes the search and carries out search-time field extractions as part of the process.

The SH applies the defined extraction rules or patterns to the raw data during the search execution. It dynamically extracts the relevant fields and associated values from the events based on these rules. The extracted fields are then used for further analysis, filtering, visualization, or other operations.

By performing search-time extractions, the SH allows users to retrieve structured and meaningful information from the raw data without modifying the underlying data source. In addition, the following are other ways in which Splunk automatically extracts fields/values during search time without user involvement:

- Name=value, also called key-value, pairs in raw data are auto-extracted
- JSON structured data is auto-extracted

Now, let's look at how you can create search-time field extractions. Here are some points to remember:

- Extractions are pre-created for every type of source in need and deployed to SHs. They are stored in config files. These extractions are reusable and work every time a user issues a search request to that source type.
- Extractions are knowledge objects. They can be shared with other users in need to promote reusability.
- The extraction approach is different for data containing delimiters and unstructured data.
- Advanced extractions require knowledge of regular expressions and deep understanding of `.conf` file (`props.conf` and `transforms.conf`) specifications.
- You can use the `rex` command in Splunk's SPL to perform extractions using regular expressions directly in the search. Knowledge of regular expressions is required to define the extraction patterns.

Let's look into extracting fields from both delimited data and unstructured data (using regular expressions) in the next sections.

Delimited data extractions

Popular structured data types that Splunk supports are CSV, **pipe-separated values (PSV)**, **tab-separated values (TSV)**, and any other data separated by a delimiter special character.

I have indexed the call record sample logs showcased in the previous section to `index=main sourcetype=callrecords`. We are going to extract the `time_of_call`, `caller`, `callee`, and `duration` fields through the Splunk Web field extractor interface. Let's begin the steps:

1. Splunk Web offers a UI for extracting delimiter-based fields. To launch the field extractor, you could search for call records events already indexed for which extractions need to be created. Expand (>) the event as shown in *Figure 11.2*.

    Then, click **Event Actions | Extract Fields**:

Figure 11.2: Splunk Web field extractor launch

Figure 11.2 shows the `callrecords` source type logs that I have indexed for demonstration. Clicking on **Extract Fields** opens the **Select Method** page, as shown in *Figure 11.3*:

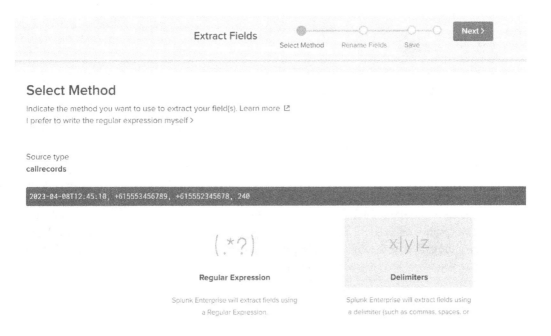

Figure 11.3: Field extractor – Select Method

Figure 11.3 shows two approaches to choose from: **Regular Expression** or **Delimiters**. The data is delimiter-based (comma-separated); hence, click on the **Delimiters** option and click **Next**. You will be taken to the page shown in *Figure 11.4*:

Figure 11.4: Field extractor – Rename Field

2. As highlighted in *Figure 11.4*, you need to choose the **Comma** option (top left). Then, the field values are auto-extracted from the sample event and assigned field names in order: **field1**, **field2**, **field3**, and **field4**. As you can see, I have renamed them (**field1** -> `time_of_call`, **field2** -> `caller`, and **field3** -> `callee`) using the pencil icon next to the field name. In the highlighted **Field Name** box, you can see that **field4** has been changed to `duration`, and clicking on the **Rename Field** button applies the changes. You can see the bottom table has the **_raw** (represents original data), **time_of_call**, **caller**, and **callee** headers applied to all the events of the **callrecords** source type. Everything so far looks fine. Clicking on the **Next** button will take you to the last step, as shown in *Figure 11.5*:

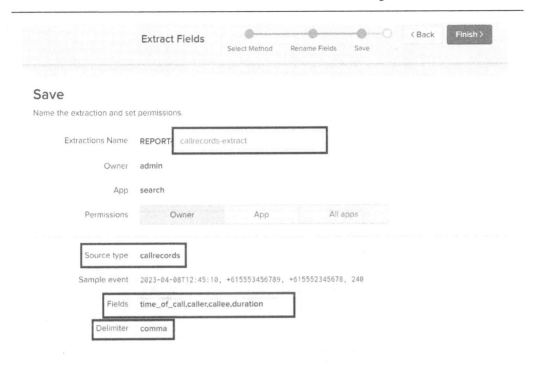

Figure 11.5: Field extractor – Save REPORT

3. *Figure 11.5* shows the final details to save; in the highlighted boxes, I have given a name to **REPORT** (`callrecords-extract`) and set **Permissions** as **Owner**, which means extraction is private to the user. The remaining **Source type**, **Fields**, and **Delimiter** areas are displaying the respective details to verify before we hit the **Finish** button. Finally, hit the **Finish** button, which will save the extraction and return a **Success!** screen.

4. Let's verify that the fields that we have created have been extracted correctly. Issue the `index=main sourcetype=callrecords` Splunk search and you will then be presented with the events. Expand (>) one of them and observe the fields and values being extracted, as shown in *Figure 11.6*:

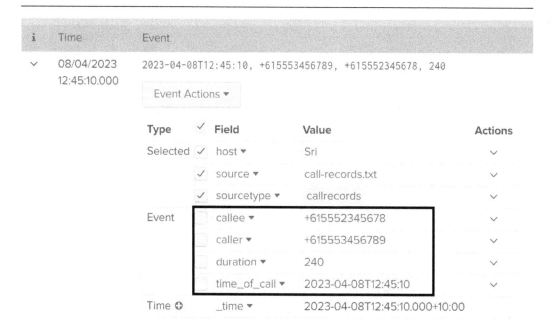

Figure 11.6: Field extractor – verification of fields

Figure 11.6 shows the four fields and values that we have created using delimiter-based extraction. In the next section, we will go through regular-expression-based field extraction from unstructured data.

> **Tip**
>
> You don't need to write search-time extractions for every source type. `splunkbase.com` is the first go-to place and contains a number of add-ons/apps that contain field extractions defined for popular technology source types, such as Microsoft, AWS, Palo Alto, Cisco, and so on.

Unstructured data extractions

The second type of search-time data extraction method that Splunk Web offers is **Regular Expression** (see *Figure 11.3*). This method is particularly suitable for data that is not structured. For example, I have considered the following web access log sample.

I ingested these logs to `index=main sourcetype=web:access:logs`:

```
127.0.0.1 - - [23/Apr/2023:10:28:12 -0400] "GET /index.html HTTP/1.1"
200 1240
127.0.0.1 - - [23/Apr/2023:10:28:13 -0400] "GET /styles.css HTTP/1.1"
200 876
127.0.0.1 - - [23/Apr/2023:10:28:15 -0400] "GET /scripts.js HTTP/1.1"
200 432
```

```
192.168.0.1 - - [23/Apr/2023:10:28:17 -0400] "GET /images/logo.png
HTTP/1.1" 200 34012
192.168.0.1 - - [23/Apr/2023:10:28:19 -0400] "GET /about.html
HTTP/1.1" 404 345
127.0.0.1 - - [23/Apr/2023:10:28:21 -0400] "GET /contact.html
HTTP/1.1" 200 2089
```

As you can see in the preceding web:access:logs source type sample, the data doesn't have delimiters and it is complex. There are some interesting fields to be captured from it – for example, to find the number of requests received from a specific IP address, how many requests are being succeeded, what type of top five HTTP methods are being issued by clients, and so on, fields are much needed.

As you can see in *Figure 11.7*, the client_ip, http_method, and http_status fields will be extracted:

Figure 11.7: Fields to be extracted – regular expressions

Let's begin the steps using the **Regular Expression** method:

1. As shown in *Figure 11.8*, search for indexed web access logs by setting **index** as **main** and **sourcetype** as **web:access:logs**. Expand (>) one of the events, and under **Event Actions**, click on **Extract Fields**:

i	Time	Event		
∨	24/04/2023 00:28:21.000	127.0.0.1 - - [23/Apr/2023:10:28:21 -0400] "GET /contact.html HTTP/1.1" 200 2089		
		Event Actions ▾		
		Build Event Type	Value	Actions
		Extract Fields	Sri	∨
		Show Source	access.log.regex	∨
		✓ sourcetype ▾	web:access:logs	∨
	Time ⊕	_time ▾	2023-04-24T00:28:21.000+10:00	
	Default	index ▾	main	∨
		linecount ▾	1	∨
		punct ▾	..._-_-_[//:::_-]_"_/_./." ___	∨
		splunk_server ▾	Sri	∨

Figure 11.8: Splunk Web field extractor launch using Regular Expression

2. Clicking on **Extract Fields** will take you to the **Select Method** screen (refer to *Figure 11.2* from the previous section). Click on **Regular Expression** and click the **Next** button. You will be presented with the **Select Fields** screen with a sample event to select the fields from, as shown in *Figure 11.9*:

Figure 11.9: Splunk Web – Regular Expression – Select Fields

3. *Figure 11.9* shows the sample event to select the required fields. As shown in *Figure 11.10*, select the portion of the sample event that represents the `client_ip`, `http_method`, and `http_status` fields and click **Add Extraction** for every field. You can see the regular expression created by Splunk by expanding **Show Regular Expression**, as highlighted in the figure:

Figure 11.10: Splunk Web – Regular Expression – selecting required fields

4. At the bottom of the same screen, **Preview** shows the fields extracted for all the events in the `web:access:logs` source type. Refer to *Figure 11.11* for the **Preview** table containing the original `_raw` event highlighted portion that's been extracted, and the `client_ip`, `http_method`, and `http_status` fields.

Figure 11.11: Splunk Web – Regular Expression – Preview

5. The preview looks perfectly fine with fields/values, so click the **Next** button to take you to the **Validate** screen. Clicking the **Next** button once again on the **Validate** screen takes you to the **Save** screen, as shown in *Figure 11.12*:

Save

Name the extraction and set permissions.

Extractions Name	**EXTRACT-** client_ip,http_method,http_status
Owner	admin
App	search
Permissions	Owner App All apps

Source type	web:access:logs
Sample event	127.0.0.1 - - [23/Apr/2023:10:28:21 -0400] "GET /contact.html HTTP/1.1" 200 2089
Fields	client_ip,http_method,http_status
Regular Expression	^(?P<client_ip>[^]+)[^"\n]*"(?P<http_method>\w+)[^"\n]*"\s+(?P<http_status>\d+)

Figure 11.12: Splunk Web – Regular Expression – Save

6. *Figure 11.12* shows the settings to be configured: **EXTRACT-** <field-names>, **Permissions** as **Owner** (private to the user), **Source type**, **Fields**, and the regular expression generated by Splunk. Finally, click the **Finish** button to save the configuration, and this will return a **Success!** screen if all goes well.

7. Let's verify the field extractions by issuing a search request, with **index** as **main** and **sourcetype** as **web:access:logs**. As shown in *Figure 11.13*, expand (>) one of the events in the search results, and you will then be able to see the fields that we have extracted using the **Regular Expression** method:

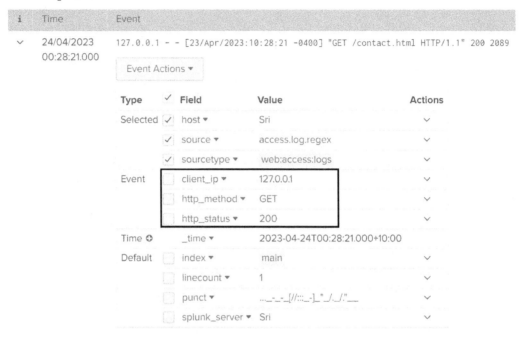

i	Time	Event			
∨	24/04/2023 00:28:21.000	127.0.0.1 - - [23/Apr/2023:10:28:21 -0400] "GET /contact.html HTTP/1.1" 200 2089			

Event Actions ▼

Type	✓	Field	Value	Actions
Selected	✓	host ▼	Sri	∨
	✓	source ▼	access.log.regex	∨
	✓	sourcetype ▼	web:access:logs	∨
Event		client_ip ▼	127.0.0.1	∨
		http_method ▼	GET	∨
		http_status ▼	200	∨
Time ⊕		_time ▼	2023-04-24T00:28:21.000+10:00	
Default		index ▼	main	∨
		linecount ▼	1	∨
		punct ▼	..._-_-_[//:::_-]_"_/._/." __	∨
		splunk_server ▼	Sri	∨

Figure 11.13: Splunk Web – Regular Expression – fields verification

That's the end of creating search-time field extractions using two methods – **Delimiters** and **Regular Expression** – through Splunk Web. The same can be accomplished using config files (`props.conf` and `transforms.conf`); however, you need a thorough understanding of the settings related to it. You can refer to Splunk docs about the direct `.conf` file approach here: `https://tinyurl.com/mrsdfwtr`.

In the next section, we will learn about creating index-time field extractions.

Creating index-time field extractions

Indexed extractions, also referred to as index-time field extractions, involve the systematic extraction of specific fields from raw data during the parsing phase of the data ingestion journey. These extractions are defined and implemented by data administrators, who specify the fields to be extracted. As part of this process, the extracted fields are not only captured but also persistently stored within the designated index, ensuring their long-term availability for subsequent analysis and retrieval.

If you recall from *Chapter 8*, in the *Data indexing phases* section, we learned about **input**, **parsing**, and **indexing**.

There is a special case for structured data: at input time, setting `INDEXED_EXTRACTIONS` in `props.conf` and deploying to a **Universal Forwarder** (**UF**) stores the fields in an index. In this case, data doesn't go through the parsing phase; it skips it and goes directly to the indexing phase. Let's look at the important facts about indexed extractions:

- Splunk, by default, stores certain fields in an index such as `index`, `source`, `sourcetype`, `host`, `splunk_server`, `linecount`, and so on. The default field list is available here: `https://tinyurl.com/53sw3xef`.

- Indexed fields are stored in the index permanently, which means they occupy additional storage space. By creating custom-indexed fields, the index size could grow significantly higher, impacting the performance of certain data sources. Search-time field extractions are preferred over indexed fields.

- Indexed fields are less flexible than search-time fields. This is because once they are written to an index, they cannot be removed unless you delete the whole index. The changes to existing indexed field extractions only apply to new data that is being ingested.

- Admins need to work with config files (`props.conf`, `transforms.conf`, and `fields.conf`) and be aware of regular expressions.

- With regards to the deployment of config files after the creation of field extractions, both `props.conf` and `transforms.conf` indexed extractions can be directed to a **Heavy Forwarder** (**HF**) if the indexers are fronted by an HF, allowing the HF to process the extractions before forwarding them to the indexers. Alternatively, if there is no HF, the extractions can be deployed directly to the indexers for processing and storage. The decision depends on the system architecture and requirements. The `fields.conf` file is deployed to the SH in a distributed deployment.

- Setting `INDEXED_EXTRACTIONS` for structured data skips the parsing phase; hence, we cannot manipulate the data, such as `index` re-routing or `sourcetype` overriding kinds of use cases.

Let's create indexed field extractions next.

Structured data extractions

If you have data in one of the following formats and you needed to index the fields, define the following setting in `props.conf` and deploy to the forwarder that is configured to monitor the file.

You could set this to a specific source type, source, and host in the following [stanza] format:

```
[<sourcetype-name> | source::<source_name> | host::<host_name>]
INDEXED_EXTRACTIONS = <CSV|TSV|PSV|W3C|JSON|HEC>
```

```
* The type of file that Splunk software should expect for a given
source
    type, and the extraction and/or parsing method that should be used
on the
    file.
* The following values are valid for 'INDEXED_EXTRACTIONS':
    CSV     - Comma-separated value format
    TSV     - Tab-separated value format
    PSV     - pipe ("|")-separated value format
    W3C     - World Wide Web Consortium (W3C) Extended Log File Format
    JSON - JavaScript Object Notation format
    HEC     - Interpret file as a stream of JSON events in the same
format as the
            HTTP Event Collector (HEC) input.

**When 'INDEXED_EXTRACTIONS = JSON' for a particular source type, do
not also
    set 'KV_MODE = json' for that source type. This causes the Splunk
software to
    extract the JSON fields twice: once at index time, and again at
search time.
```

With this setting, the forwarder assumes file types of CSV, TSV, and PSV having a header as the first line in a file. If there is no header present in a file, setting FIELD_NAMES is required to name the suitable fields in order.

For example, consider the callrecords example that we saw in the previous search-time extractions section.

Here are the CSV callrecords logs with a header:

```
time_of_call,caller,callee,duration
2023-04-08T09:30:15, +615551234567, +615552345678, 120
2023-04-08T10:15:42, +615551234567, +615553456789, 60
2023-04-08T11:20:30, +615552345678, +15551234567, 180
2023-04-08T12:45:10, +615553456789, +615552345678, 240
```

The following extraction setting applies to the callrecords source type. It will permanently store the time_of_call, caller, callee, and duration indexed fields along with values in the designated index:

```
### props.conf for callrecords source type
### deploy to a forwarder monitoring CSV file with header
[callrecords]
INDEXED_EXTRACTIONS = CSV
```

If you are interested in reading more about various other settings that are useful in cases such as where files have no header, customizing timestamp extraction, non-standard delimiter cases, and more, refer to Splunk docs (https://tinyurl.com/32usnknz). These settings are not necessarily going to be tested in the Admin Certification exam.

In the next section, we will look into creating custom-indexed fields for unstructured data.

Unstructured data extractions

To create indexed filed extractions, we are once again using the same web:access:logs sample that we saw in the search-time extractions section:

```
127.0.0.1 - - [23/Apr/2023:10:28:12 -0400] "GET /index.html HTTP/1.1"
200 1240
127.0.0.1 - - [23/Apr/2023:10:28:13 -0400] "GET /styles.css HTTP/1.1"
200 876
127.0.0.1 - - [23/Apr/2023:10:28:15 -0400] "GET /scripts.js HTTP/1.1"
200 432
192.168.0.1 - - [23/Apr/2023:10:28:17 -0400] "GET /images/logo.png
HTTP/1.1" 200 34012
192.168.0.1 - - [23/Apr/2023:10:28:19 -0400] "GET /about.html
HTTP/1.1" 404 345
127.0.0.1 - - [23/Apr/2023:10:28:21 -0400] "GET /contact.html
HTTP/1.1" 200 2089
```

The fields that we are going to extract here are client_ip, http_method, and http_status. Refer to *Figure 11.7* to identify the portion of text being considered for field values.

High-level steps that are involved to create index-time field settings are described as follows:

1. Define props.conf referring to the transforms.conf stanza name:

```
#### props.conf specification
[<sourcetype_name> | source::<source-name> | host::<host-name> ]
TRANSFORMS-<classname> = <unique_transforms_stanza_name>
```

2. Define transforms.conf with the field extraction settings:

```
## transforms.conf specification
[<unique_transforms_stanza_name>]
REGEX = <regular_expression>
FORMAT = <your_custom_field_name>::$1
WRITE_META = [true|false]
DEST_KEY = <KEY>
DEFAULT_VALUE = <string>
SOURCE_KEY = <KEY>
REPEAT_MATCH = [true|false]
LOOKAHEAD = <integer>
```

3. Define `fields.conf`:

    ```
    ## fields.conf specification
    [<field_name>]
    INDEXED = true
    ```

4. Package `props.conf` and `transforms.conf` into an app and deploy to the HF/indexer in a distributed deployment. Deploy `fields.conf` to the SH in a distributed deployment.

5. In the case of a standalone Splunk instance, all `.conf` files are deployed to the same single instance and a restart is required afterward.

We have understood the specification of required `.conf` files and settings, now let's get into creating these settings for the `web:access:logs` source type example:

1. Create the `props.conf` file for the `sourcetype` stanza:

    ```
    [web:access:logs]
    TRANSFORMS-web_extractions = web_access_idx_extractions
    ```

2. Create `transforms.conf`:

    ```
    [web_access_idx_extractions]
    SOURCE_KEY = _raw
    REGEX = ^([^ ]+)[^"\n]*"(\w+)[^"\n]*"\s+(\d+)
    FORMAT = client_ip::$1 http_method::$2 http_status::$3
    WRITE_META = true
    ```

3. Create `fields.conf`:

    ```
    [client_ip]
    INDEXED = true

    [http_method]
    INDEXED = true

    [http_status]
    INDEXED = true
    ```

That's it, we have completed the required configurations. Deploying them to the respective HF/indexer and `fields.conf` to the SH will start creating indexed fields for the newly ingested data.

Creating lookups

We explored what a lookup is and some of its types in the *Understanding fields and lookups* section. Lookups in Splunk are crucial for enriching and correlating data, enabling efficient analysis and advanced search capabilities. In this section, we are going to create CSV and KV Store lookup files using Splunk Web.

As an example, we will use a lookup of country codes to country names as follows. If you recall the `callrecords` sample from the previous section, the data contains phone numbers with country codes, but we can't find out the origin country of the phone numbers from `callrecords` Splunk events alone. In order to obtain the country name from the country code, the knowledge managers or data administrators create additional lookups. The lookup can be further used in Splunk queries to correlate country codes with the lookup and retrieve country names from it.

Save the following contents in a file as `phone_no_country_code_to_name.csv` in your local system:

```
country_code, country_name
+61,Australia
+91,India
+1,USA
+47,Norway
+358,Finland
+33,France
+44,UK
```

CSV lookups

There are two major steps involved in creating lookups:

- Creating a lookup table
- Creating a lookup definition

Let's get into the steps of creation:

1. Log in to Splunk Web and go to the **Settings** menu, click **Lookups**, and click **Lookup table files**.
2. Clicking on the **New Lookup Table File** button in the top-right corner will open the following page (see *Figure 11.14*):

Figure 11.14: Lookup table files – Add new

3.   *Figure 11.14* shows the **Destination app** option as **search**. I clicked on the **Choose File** button and selected the file that I have saved in the local system and entered `phone_no_country_code_to_name.csv` for **Destination filename**. Finally, click **Save**. It will take you to a page with a **Successfully saved** message, as shown in *Figure 11.15*:

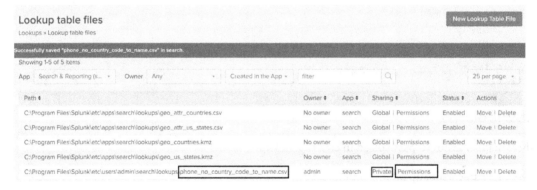

Figure 11.15: Lookup table files – saved successfully

4.   *Figure 11.15* shows the lookup file uploaded, with **Sharing** set as **Private** and having the **Permissions** option to share with other users.

5.   Let's create a lookup definition. Go to **Settings**, click **Lookups**, click **Lookup definitions**, and then click on **New Lookup Definition**. A new page will be displayed, as shown in *Figure 11.16*:

Destination app	search	▾
Name *	phone_no_country_code_to_name	
Type	File-based	▾
Lookup file *	phone_no_country_code_to_name.csv	▾

Create and manage lookup table files.

☐ Configure time-based lookup

☐ Advanced options

Cancel   Save

Figure 11.16: Lookup definition – Add new

6. *Figure 11.16* shows **Destination app** as **search**, the name entered for the definition as phone_no_country_code_to_name, **Type** is by default populated as **File-based**, and from the **Lookup file** dropdown, I have selected the lookup table file that we uploaded in *step 3*. Finally, clicking **Save** will take you to the page with a **Successfully saved** message:

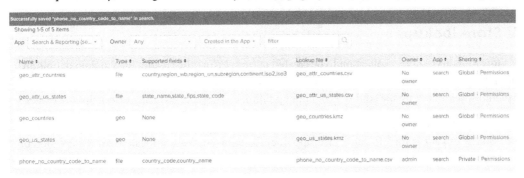

Name ⬍	Type ⬍	Supported fields ⬍	Lookup file ⬍	Owner ⬍	App ⬍	Sharing ⬍
geo_attr_countries	file	country,region_wb,region_un,subregion,continent,iso2,iso3	geo_attr_countries.csv	No owner	search	Global ⏐ Permissions
geo_attr_us_states	file	state_name,state_fips,state_code	geo_attr_us_states.csv	No owner	search	Global ⏐ Permissions
geo_countries	geo	None	geo_countries.kmz	No owner	search	Global ⏐ Permissions
geo_us_states	geo	None	geo_us_states.kmz	No owner	search	Global ⏐ Permissions
phone_no_country_code_to_name	file	country_code,country_name	phone_no_country_code_to_name.csv	admin	search	Private ⏐ Permissions

Figure 11.17: Lookup definition saved successfully

7. *Figure 11.17* shows the saved definition (phone_no_country_code_to_name) in the **search** app, and **Sharing** is set as **Private** with the **Permissions** option to share with other users.

That's it, we have successfully created a lookup in Splunk. To verify the work we have done so far, open a search window under the search app and issue the | inputlookup phone_no_country_code_to_name query. As shown in *Figure 11.18*, you will be presented with country_code and country_name in tabular format:

In the next section, we will learn about KV Store lookups.

KV Store lookups

KV Store lookups require a collection to be created upfront before the creation of a lookup definition. Splunk Web doesn't offer the creation of collections through the UI. Instead, a Splunk-supported app – Splunk App for Lookup file editing (aka Lookup Editor) – at `splunkbase.com` helps to create it.

Follow the installation steps for the app at `https://splunkbase.splunk.com/app/1724` and install it in your Splunk instance. The app does the following high-level steps for the creation of KV Store lookups:

- Creates a collection (`collections.conf`)
- Creates a lookup definition

We are using the same `country_code, country_name` lookup that we created in the previous section for KV Store as well:

1. After installation, launch **Splunk App for Lookup File Editing** from the **Apps** menu. In the top-right corner, click on **Create a New Lookup** | **KV Store Lookup**. *Figure 11.19* shows the options. It will open a page to create a new lookup:

Figure 11.19: New KV Store lookup – Lookup Editor

2. *Figure 11.19* shows the name to be given for the collection, the app, and **Key-value collection schema** details. I entered kv_country_code_to_name in **Name**; for **App,** I selected **Search & Reporting** from the dropdown; and for **Field**, I entered country_code and country_name. Finally, click the **Create Lookup** button at the bottom right of the screen.

| Lookups | New Lookup | Health ▾ | Search ▾ |

Lookups / New Lookup

Name: `kv_country_code_to_na`   App: `Search & Reporting` ▾   ☐ Replicate ❓

Specifies the name of collection   Specifies the app where the collection will reside

Key-value collection schema

| Field: | country_code | String | ∨ | Remove |

| Field: | country_name | String | ∨ | Remove |

| Field: | Name | String | ∨ | Remove |

| Field: | Name | String | ∨ | Remove |

Figure 11.20: New Lookup – create lookup

3. An empty collection with a lookup definition is created by the Lookup Editor app. You can add the rows accordingly, as shown in *Figure 11.21*:

Figure 11.21: Lookup collection – created

4. To access the collection from search, a `transforms` stanza needs to be created. To do that, click on the **Lookups** menu item in the top-left corner of *Figure 11.22*:

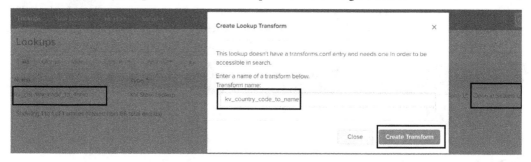

Figure 11.22: Lookups – Create Transform

5. As shown in *Figure 11.22*, search for the `kv_country_code_to_name` lookup that we created in previous steps and click on the **Open in Search** link. This will open a **Create Lookup Transform** popup. In the **Transform name** textbox, enter `kv_country_code_to_name` and click the **Create Transform** button.

6. Now, open a search bar under the Search App and issue a `| inputlookup append=t kv_country_code_to_name` query.

    As shown in *Figure 11.23*, you will be presented with the results that we created in *step 3*:

Figure 11.23: KV Store lookup – final verification

We have successfully created a KV Store lookup using the Lookup Editor app. Without the app, we would have to create the collection through the `collections.conf` file and define transforms to access the lookup in search. The app is really handy, isn't it? It is capable of creating CSV lookups as well so that it creates both CSV table files and table definitions through simple steps rather than switching to multiple pages. If you want to give it a try, you could do it by referring to *Figure 11.18*, where you will find the **CSV lookup** option, and follow the on-screen steps.

It is highly recommended that you understand the management of permissions for knowledge objects to enable you to share knowledge objects such as field extractions and lookups with other users. The sharing of objects can be restricted to specific roles and apps and can be set to **All Apps (System)**. There are a variety of permission combinations that can be applied, which depend on individual Splunk deployment scenarios. "Knowledge object permissions" is not an exam topic; however, I strongly recommend you go through the Splunk documentation to learn more: `https://tinyurl.com/79h4mrrw`.

That is everything you need to know about lookups for the certification exam. We have come to the end of the chapter. Let's summarize what we have learned in the next section.

Summary

The chapter combines two interesting topics: fields and lookups. We began with fields (or knowledge objects) and understood that they are the building blocks for search (SPL) and that they can be extracted out of raw data while presenting results to search requests issued by users during search time. The SH, on the fly, extracts fields/values based on the extraction's settings pre-created by data admins on the SH. Splunk, by default, extracts fields from data formats such as KV pairs and JSON contents during search time. Then, we introduced lookup types and their purpose for data enrichment use cases.

Similarly, another approach is to define the fields through indexed extraction settings during data indexing time; they are called index-time fields/indexed fields. Fields created during indexing time are stored in a designated index permanently, so they consume additional storage space, and using them is a less preferred approach over using search-time fields. The reason for this is that indexed fields are less flexible and might impact search performance. Certain data formats (such as CSV, PSV, TSV, and so on) could be enabled to extract fields in the input phase on forwarders by setting `INDEXED_EXTRACTIONS`; in this case, the parsing phase is skipped.

We have created both search-time field extractions through Splunk Web and indexed fields through `.conf` files (`props.conf`, `transforms.conf`, and `fields.conf`). Splunk Web offers two approaches for search-time field extractions; they are **Delimiter** for formatted data and **Regular Expression** for unstructured data. We understood the deployment side of it: for indexed extractions, the `props.conf` and `transforms.conf` files will be deployed to the HF/indexer, and `fields.conf` to an SH in a distributed deployment.

Finally, we created CSV lookups through Splunk Web by creating a lookup table file and a lookup definition to access it in search. We installed the Lookup Editor app for KV Store lookup file creation and created transforms to access the lookups in the search. We understood that the app can create both CSV and KV Store lookups much easier than the standard Splunk Web approach.

I would like to take a moment to express my heartfelt gratitude to you for joining me on this journey through the pages of this book. Exploring the intricacies of Splunk and its various features has been a pleasure, and I hope you have found value and insight within these chapters.

As you continue your Splunk journey, I want to wish you all the best in your pursuit of the Splunk Certified Admin exam. Remember to stay focused, review the material thoroughly, and practice the hands-on exercises to strengthen your skills.

Remember that learning is a continuous process, and there is always more to discover. I encourage you to continue exploring and expanding your understanding of Splunk, leveraging its capabilities to unlock new insights and solutions. Once again, thank you for accompanying me on this journey. May your endeavors in the realm of Splunk be filled with success and meaningful discoveries.

In the next chapter, you will be provided with a mock exam for your self-assessment in the actual exam format of 56 questions. We will also discuss the answers at the end to check your attempts and refresh your understanding. In the next section, you have been given sample questions followed by answers to test your learning about Splunk fields and lookups.

Self-assessment

Like other chapters, this section is about self-testing your knowledge of the topics that we have learned about so far. You have 10 questions, and the answers are given after the questions. As reiterated every time, feel free to review the respective sections once again if you have difficulty answering questions. Let's begin:

1. Are field names case-sensitive?

    A. Yes

    B. No

2. Does Splunk auto-extract fields/values from the key-value format in search time?

    A. Yes

    B. No

3. Select the types of lookup files available in Splunk (select all that apply):

    A. JSON lookups

    B. CSV lookups

    C. KV Store lookups

    D. Geospatial lookups

4. Which statements are true about CSV lookups? Select all that apply:

   A. CSV lookups store information in MongoDB

   B. CSV lookup files have the .csv extension

   C. CSV lookup files are used to enrich information by correlating with existing events in Splunk.

   D. CSV lookups contain information in comma-separated values with a header

5. What are the two types of Splunk Web methods available for search-time field extraction creation?

   A. Regular Expression

   B. KV pair extraction

   C. Delimiter extraction

   D. JSON default extraction

6. Does the Regular Expression method in Splunk Web generate a regular expression by itself?

   A. Yes

   B. No

7. Is the storage of index-time fields permanent in Splunk's index?

   A. Yes

   B. No

8. Identify the default fields that Splunk extracts and stores during the indexing phase (select all that apply):

   A. source

   B. sourcetype

   C. host

   D. splunk_server

9. Can we set INDEXED_EXTRACTIONS on the UF for structured contents as CSV and PSV types?

   A. No

   B. Yes

10. Does creating custom index-time field extractions occupy additional storage space on the indexer?

    A. No

    B. Yes

Reviewing answers

1. *Option A* is correct. Field names are case-sensitive. For example, `http_status` and `HTTP_STATUS` are seen as two different fields by Splunk.

2. *Option A* is correct. Splunk auto-extracts field names/values from data containing the KV format and JSON type format.

3. *Options B, C, and D* are correct. One more lookup type exists, which is called external lookups.

4. *Options B, C, and D* are correct. KV Store lookups are backed by the MongoDB engine. CSV lookups are stored in the SH filesystem with the `.csv` extension.

5. *Options A and C* are correct. Only two methods are available in Splunk Web.

6. *Option A* is correct. The Splunk Web Regular Expression method generates the regular expression by itself. This method doesn't require extensive knowledge of regular expressions to extract search-time fields.

7. *Option A* is correct. Index-time extraction fields/values are stored in the designated index permanently.

8. *All options* are correct. Default fields such as `splunk_server`, `source`, `sourcetype`, `host`, and `index` are indexed fields. There are other indexed fields, called internal fields, which are prefixed with `_` such as `_indextime`, `_time`, `_raw`, `_cd`, and so on.

9. *Option B* is correct. `INDEXED_EXTRACTION` works on the forwarder running on the source. Yet if it was a UF, then it must be set there in the `props.conf` file. Setting this one on the UF skips the parsing phase and all the values in structured data will be stated indexed field/values.

10. *Option B* is right. Creating custom indexed-field extractions occupies additional space on indexers. Indexed fields are less flexible as they are stored permanently in the designated index. Search-time fields are preferred over indexed fields.

Self-Assessment Mock Exam

Welcome to the mock exam chapter of the *Splunk Enterprise Certified Administration Guide*! Congratulations on reaching this milestone in your journey to becoming a certified Splunk Enterprise admin. In this chapter, we will delve into a series of self-assessment questions and answers designed to test your knowledge and reinforce the concepts covered throughout the book.

Before we proceed, I want to commend you for your dedication and hard work in preparing for this exam. You have gained a comprehensive understanding of the Splunk platform, its administration, and its powerful features. This final chapter will provide an opportunity for you to assess your progress and identify any areas that may require further study or clarification. Remember—the self-assessment questions are meant to challenge your understanding of Splunk Enterprise administration concepts, so approach them with an open mind and utilize the knowledge you have acquired throughout your studies. Take your time, read the questions carefully, and select the most appropriate answer.

Once you have completed the self-assessment questions, you can refer back to the answers provided in this chapter to evaluate your performance. It's essential to review any incorrect answers and understand the underlying concepts to enhance your knowledge further. Keep in mind that this self-assessment is not intended to mimic the actual certification exam. Instead, it serves as a tool to gauge your preparedness and highlight any potential areas of improvement. Use it as an opportunity to fine-tune your understanding and solidify your knowledge.

The original exam duration is 60 minutes for 56 questions and requires you to manage your time effectively. It's important to allocate approximately 1 minute to each question, considering the time you will need to read and understand each question.

Here's a breakdown of the suggested time management for the exam:

- **Reading, understanding, and answering each question**: 1-1.5 minutes per question
- **Reviewing your answers**: 3 minutes at the end of the exam

By adhering to this time management strategy, you can ensure that you have sufficient time to read and comprehend each question, provide accurate answers, and review your responses before submitting the exam.

For this self-assessment chapter, you will be given a total of 56 questions, just as in the original exam format. These questions have been carefully designed to cover various aspects of Splunk Enterprise administration, ensuring a comprehensive assessment of your knowledge and skills. Remember—confidence comes from preparation. So, embrace this final chapter as a chance to reinforce your expertise and approach the certification exam with certainty. Best of luck on your journey to becoming a certified Splunk Enterprise admin, and let's get started with the self-assessment questions!

Mock exam questions

1. Which configuration file in Splunk is responsible for specifying data inputs to be collected and indexed?

    A. `inputs.conf`

    B. `props.conf`

    C. `transforms.conf`

    D. `indexes.conf`

2. You are a system administrator: how can you control access to specific indexes and resources in Splunk?

    A. By configuring firewall rules on the Splunk servers.

    B. By encrypting the data before indexing it in Splunk.

    C. By setting up authentication mechanisms such as the **Lightweight Directory Access Protocol (LDAP)** or **Security Assertion Markup Language (SAML)**.

    D. By following Splunk **role-based access control (RBAC)** and creating a role that can be configured to allow access to specific indexes and resources. Roles in turn can be assigned to users.

3. Which Splunk component is responsible for deploying apps to forwarders?

    A. The **deployment server (DS)**

    B. The **heavy forwarder (HF)**

    C. The **license manager (LM)**

    D. The **search head (SH)**

4. What is the purpose of a Splunk indexer?

    A. It indexes and stores incoming data for searching and analysis

    B. It provides a web-based interface for searching and visualizing data

    C. It manages user authentication and access control

    D. It runs scheduled searches and generates alerts based on defined criteria

5. Which authentication method allows users to log in to Splunk Enterprise using their existing LDAP or **Active Directory** (**AD**) credentials?

    A. Internal/native authentication

    B. SAML authentication

    C. LDAP authentication

    D. External authentication

6. Which component of Splunk Enterprise is responsible for running searches and scheduling alerts?

    A. The indexer

    B. The deployment server

    C. The search head

    D. The license manager

7. Can Splunk perform analysis on binary data?

    A. Yes—Splunk can analyze binary data using custom parsers and extraction methods

    B. No—Splunk is limited to analyzing only text-based data

8. What happens when a Splunk Enterprise license is in violation?

    A. Splunk stops ingesting new data but retains access to existing data

    B. Splunk continues to function normally without any limitations

    C. Splunk restricts access to search functionality but retains indexing capabilities

    D. Splunk disables all data collection and search functionality

9. How can you monitor and manage your Splunk license usage?

    A. Splunk provides built-in monitoring and reporting dashboards for license usage

    B. License usage can only be monitored through manual data analysis

    C. Splunk requires additional apps for license monitoring

    D. License usage can only be managed through Splunk support services

10. What are the different types of Splunk licenses? (Select all that apply)

    A. A **Free** license, a **Trial** license, and **Enterprise** licenses

    B. **Basic**, **Standard**, and **Premium** licenses

    C. A Splunk Enterprise infrastructure license

    D. Splunk licenses are customized based on specific user requirements

11. What is a license violation in Splunk?

    A. A license violation occurs when the Splunk software is not installed correctly

    B. A license violation happens when data exceeds the licensed volume or other usage limits

    C. A license violation refers to unauthorized access to Splunk's licensing server

    D. A license violation means using outdated or expired license keys

12. Can license stacking and license pools be used together in Splunk?

    A. Yes—license stacking and license pools can be combined for flexible license management

    B. No—license stacking and license pools are mutually exclusive features in Splunk

13. What is the role of the LM component in Splunk?

    A. The LM is responsible for renewing Splunk licenses

    B. The LM distributes and assigns license keys to other components

    C. The LM is a repository of licenses in a distributed deployment

    D. The LM is a third-party software tool used to manage licenses for Splunk

14. What is the default Splunk web port?

    A. 8089

    B. 8000

    C. 8191

    D. 9997

15. Roles in Splunk determine which actions a user can perform within the platform.

    A. True

    B. False

16. How can users be added to Splunk?

    A. Using the Splunk web interface only

    B. Using the **command-line interface (CLI)** only

    C. Both the Splunk web interface and the CLI

    D. None of these

17. What does user role inheritance mean in Splunk?

  A. Users inherit the capabilities and indexes only of the roles assigned to them

  B. Users inherit the roles of other users in their group

  C. Roles inherit the capabilities and indexes of other roles

  D. Roles inherit the authentication methods of other roles

18. What are the default roles in Splunk?

  A. `admin`, `power`, `can_delete`, and `user`

  B. Splunk Admin, Splunk User, and Splunk Analyst

  C. Splunk Admin, Splunk Power User, and Splunk User

  D. Administrator, Analyst, and Operator

19. How can you manage user roles and permissions in Splunk?

  A. Through the Splunk web interface under **Settings | Roles**

  B. By modifying the `authorize.conf` file on the Splunk SH

  C. It is not possible to modify default roles in Splunk

  D. By contacting Splunk support for assistance with role management

20. What is the CLI command for adding an indexer to a forwarder?

  A. `./splunk add forward-server <server:port>`

  B. `./splunk add indexer <server:port>`

  C. `./splunk add config-forwarder <server:port>`

  D. `./splunk edit forwarder-config <server:port>`

21. Which component is responsible for sending data from forwarders to the Splunk indexers?

  A. The Deployment Server

  B. The **cluster manager** (**CM**)

  C. The Search Head

  D. The forwarder itself

22. What is the default phone home interval on **deployment clients** (**DCs**)?

  A. 60 seconds

  B. 90 seconds

  C. 120 seconds

  D. 10 seconds

23. How can you configure a DC to communicate with a DS in Splunk? (Select all that apply)

   A. By specifying the DS's address in the `deploymentclient.conf` file

   B. By using the Splunk web interface at **Settings | Deployment Client**

   C. By executing a CLI command on the DC

   D. By contacting Splunk support for assistance with the configuration

24. Can you identify the config file in which index creation settings are defined?

   A. `server.conf`

   B. `inputs.conf`

   C. `indexes.conf`

   D. `outputs.conf`

25. What is the location of both hot and warm buckets stored in a Splunk indexer? (Select the config setting)

   A. `hotPath`

   B. `homePath`

   C. `coldPath`

   D. `warmPath`

26. What is the setting that defines the max size of an index?

   A. `maxTotalDataSizeMB`

   B. `maxDataSizeMB`

   C. `indexMaxSize`

   D. `indexTotalSize`

27. What are the commands to delete data from an index? (Select all that apply)

   A. The `delete` command through a Splunk search request

   B. The `delete` command through the Splunk CLI

   C. The `splunk clean` command through a Splunk search request

   D. The `splunk clean` command through the Splunk CLI

28. Can Splunk search the frozen buckets of an index?

   A. Yes

   B. No

29. What are the index settings that are useful to roll over cold buckets to frozen buckets without permanently deleting the data? (Select all that apply)

    A. `frozenPath`

    B. `coldToFrozenDir`

    C. `coldPath`

    D. `coldToFrozenScript`

30. What is the main use of a thawed bucket in Splunk?

    A. To untar index files

    B. To roll over cold buckets for archiving

    C. To restore frozen buckets and make them searchable

    D. To block specific data participating in searches

31. In which config files are source types defined in Splunk?

    A. `transforms.conf`

    B. `sourcetypes.conf`

    C. `props.conf`

    D. `server.conf`

32. Can you identify the directory locations on Splunk where config files are usually present? (Select all that apply)

    A. `$SPLUNK_HOME/bin`

    B. `$SPLUNK_HOME/etc/users/<user_name>/<app_name>/[default|local]`

    C. `$SPLUNK_HOME/etc/apps/<app_name>/[default|local]`

    D. `$SPLUNK_HOME/etc/system/[default|local]`

33. What is the top-level directory for search-time (app/user) precedence in Splunk?

    A. `$SPLUNK_HOME/etc/system/local`

    B. `$SPLUNK_HOME/etc/apps/<app_name>/default`

    C. `$SPLUNK_HOME/etc/users/<app_name>/local`

    D. `$SPLUNK_HOME/etc/system/default`

34. What is the top-level directory for index-time (global) precedence in Splunk?

    A. `$SPLUNK_HOME/etc/apps/<app_name>/default`

    B. `$SPLUNK_HOME/etc/users/<app_name>/local`

    C. `$SPLUNK_HOME/etc/system/default`

    D. `$SPLUNK_HOME/etc/system/local`

35. What is the location in which Splunk system-wide configurations by default are stored?

    A. `$SPLUNK_HOME/etc/system/default`

    B. `$SPLUNK_HOME/etc/apps/`

    C. `$SPLUNK_HOME/bin`

    D. `$SPLUNK_HOME/etc/apps/`

36. What is the minimum number of SHs that are required in an **SH cluster (SHC)**?

    A. Two

    B. Three

    C. Four

    D. One

37. The **search factor (SF)** is configured on which Splunk component?

    A. The SH

    B. The indexer

    C. The SH captain

    D. The LM

38. If a **replication factor (RF)** of 4 has been set on an indexer cluster, what is the minimum number of indexer instances that are required in a cluster?

    A. Three

    B. Two

    C. Four

    D. Eight

39. Can you provide the names of the two configuration files that contain the distributed search settings on the SH?

    A. `outputs.conf`

    B. `server.conf`

    C. `serverclass.conf`

    D. `distsearch.conf`

40. What is the default `replicationPolicy` value for knowledge bundle replication on the SH?

    A. `replicationPolicy = cascading`

    B. `replicationPolicy = classic`

    C. `replicationPolicy = mounted`

    D. `replicationPolicy = rfs`

41. What is the default `maxBundleSize` limit for knowledge bundle replication in the SH?

    A. 2 GB

    B. 4 GB

    C. 1 GB

    D. 3 GB

42. During the ingestion process in Splunk, data undergoes three phases. Which of the following are these three phases of data processing in Splunk?

    A. Input

    B. Parsing

    C. Indexing

    D. Replication

43. Where should the configuration for file/directory monitoring be included in Splunk?

    A. `transforms.conf`

    B. `inputs.conf`

    C. `props.conf`

    D. `indexes.conf`

44. What is the Splunk CLI command used to add inputs in Splunk?

    A. `splunk add monitor <path_to_data_source> -index <index_name> -sourcetype <sourcetype_name>`

    B. `splunk add inputs <path_to_data_source> -index <index_name> -sourcetype <sourcetype_name>`

    C. `splunk add files <path_to_data_source> -index <index_name> -sourcetype <sourcetype_name>`

    D. `splunk add directory <path_to_data_source> -index <index_name> -sourcetype <sourcetype_name>`

45. What is the **HTTP Event Collector** (**HEC**) endpoint that allows ingestion of any text payload in a request?

    A. `services/collector/event`

    B. `services/collector/raw`

    C. `services/collector/text`

    D. `services/collector/ack`

46. How can you selectively ingest specific event codes for Windows inputs in Splunk?

    A. By using `Event_codes = <comma separated codes list>` in `inputs.conf`

    B. By using whitelist/blacklist settings in `inputs.conf`

    C. There is no possible solution to selectively ingest, so filter the codes while searching

    D. None of these

47. Can you identify the data inputs through which data is ingested into Splunk? (Select all that apply)

    A. The HEC—agentless input

    B. File and directory monitoring

    C. Network input

    D. By using **technology add-ons** (**TAs**) for data collection

48. How do you monitor sub-directories recursively in file and directory input?

    A. `*` (wildcard)

    B. `...` (three-dot notation ellipsis in a directory segment)

    C. By configuring individual sub-directories as a monitoring input

    D. `%%` (using this notation in directory segments)

49. What are the conf files that are required to configure parsing phase settings? (Select all that apply)

   A. `props.conf`

   B. `server.conf`

   C. `transforms.conf`

   D. `limits.conf`

50. What is the setting that is useful for forwarding data to a selected indexer `tcpout` group in `inputs.conf`?

   A. `IndexerGroup`

   B. `_TCP_ROUTING`

   C. `defaultIndexer`

   D. `indexerList`

51. Select the two Splunk methods that are available in Splunk Web for search-time field extractions.

   A. Delimiters

   B. Structured data extractions

   C. **Regular expressions (regex)**

   D. Unstructured data extractions

52. Do index-time field extractions store fields in a designated index?

   A. No

   B. Yes

53. In a distributed deployment, index-time field extractions and search-time field extractions are deployed to which components?

   A. Index-time fields: the SH; search-time fields: indexers

   B. Index-time fields: indexers; search-time fields: the SH

   C. Index-time fields: CM; search-time fields: SH deployer

   D. Index-time fields: the SH; search-time fields: the **universal forwarder** (UF)

54. Can you select the correct lookup read commands? (Select all that apply)

   A. `lookup`

   B. `outputlookup`

   C. `inputlookup`

   D. `readlookup`

55. As part of data ingestion, which phase is skipped from execution by setting `INDEXED_EXTRACTIONS` on a forwarder?

    A. Input

    B. Parsing

    C. Indexing

    D. Ingestion

56. Why is a persistent queue useful in Splunk?

    A. It helps reduce the overall storage requirements in Splunk

    B. It enables real-time alerting and notification for critical events

    C. It prevents data loss during high-volume data ingestion

    D. It improves query performance for historical data analysis

You have come to the end of this *Mock exam questions* section. Let's review the answers.

Reviewing answers

1. *Option A* is the correct answer. The `inputs.conf` file typically contains the data inputs for ingestion. `props.conf` and `transforms.conf` contains parsing phase configurations and field extractions, and `indexes.conf` configures Splunk indexes. We briefly went through these files in *Chapter 6, Splunk Configuration Files*.

2. *Option D* is the correct answer. *Option C* is about authentication methods that are required for user authentication to Splunk. Refer to the *Roles* and *Authentication methods* sections of *Chapter 3, Users, Roles, and Authentication in Splunk*.

3. *Option A* is the correct answer. The DS stores applications for forwarders that are connected to it. By using the `serverclass.conf` configuration file, the DS configures the necessary applications to be deployed to the forwarders. Refer to the *Understanding Splunk components* section of *Chapter 1, Getting Started with the Splunk Enterprise Certified Admin Exam*, for greater detail about various Splunk components.

4. *Option A* is the correct answer. The important function of indexers is to store data and respond to search requests issued by users via the SH. In a typical production deployment, the web UI is disabled on indexers. *Options B, C*, and *D* are features of the SH.

5. *Option C* is the correct answer. As the name implies, the LDAP authentication method is typically configured to connect to the organization's existing directory server and let users authenticate using existing credentials. Other authentication methods, such as SAML, native, and external, are supported by Splunk.

6. *Option C* is the correct answer. The SH exposes various interfaces to users to issue ad hoc search requests and schedule searches. The remaining indexers, DS, and LM are Splunk components dedicated to other purposes in Splunk distributed deployment.

7. *Option B* is correct. Splunk can only process text data.

8. *Option C* is the correct answer. Splunk stops the search feature and continues indexing data within a violation period.

9. *Option A* is the correct answer. The **monitoring console (MC)** has built-in dashboards/reports for monitoring license usage.

10. *Options A* and *C* are the correct answers. The remaining options don't even exist in Splunk. Refer to the *Introducing license types* section of *Chapter 2, Splunk License Management*, for all available license types.

11. *Option B* is the correct answer.

12. *Option A* is the correct answer. License stacking is possible only with Splunk Enterprise licenses and Splunk Enterprise Trial licenses. License pools can be created out of these stacks. Refer to the *License groups, stacks, and pools* section of *Chapter 2, Splunk License Management*.

13. *Option C* is the correct answer. The LM is a Splunk instance where licenses are being installed and to which all Splunk instances (alias license peers) in a distributed deployment are connected for license information.

14. *Option B* is correct. `8000` is the default web port, `8089` is the management port, `9997` is the receiving port, and `8191` is a **key-value (KV)** store replication port.

15. *Option A* is correct. A role contains the necessary permissions and capabilities a user has, along with resource limits.

16. *Option C* is correct. Splunk offers four standard interfaces—Splunk Web, the Splunk CLI, a RESTful API, and the direct creation or update of the necessary config file.

17. *Option C* is correct. A role can optionally inherit other roles, and users are assigned to roles. Refer to the *Roles* section of *Chapter 3, Users, Roles, and Authentication in Splunk*.

18. *Option A* is correct. There are no such roles in Splunk as given in *options B, C*, and *D*.

19. *Options A* and *B* are correct. *Options C* and *D* are incorrect: default roles can be modified; however, this is not advisable. Splunk support is only able to help with Splunk product issues.

20. *Option A* is correct. Refer to *Chapter 4, Splunk Forwarder Management*.

21. *Option D* is correct.

22. *Option A* is correct. The default phone home interval on the forwarder is 60 seconds, which can be updated if required in `deploymentclient.conf` on the forwarder.

23. *Options A* and *C* are correct. The CLI command is `./splunk set deploy-poll <ds-servername:management_port>`.

24. *Option C* is correct. The indexes.conf file contains Splunk index details. Refer to *Chapter 5, Splunk Index Management.*

25. *Option B* is correct. There is no such setting as hotPath.

26. *Option A* is correct. There are no such settings as indexMaxSize or indexTotalSize.

27. *Options A* and *D* are correct. With the delete command, you can filter the data that you want to delete. splunk clean eventdata -index <index_name> is a CLI command that removes all data in the index permanently from disk with no filtering options. To find more details about the clean command on a Splunk instance issue, execute the splunk help clean command.

28. *Option B.* No—Splunk won't read frozen buckets; they must be thawed and rebuilt for searching. Refer to the *Understanding buckets* section of *Chapter 5, Splunk Index Management.*

29. *Options B* and *D* are the correct answers. There is no such setting as frozenPath.

30. *Option C* is correct.

31. *Option C* is correct.

32. *Options B, C,* and *D* are the correct answers. Refer to the *Understanding conf file precedence* section of *Chapter 6, Splunk Configuration Files.*

33. *Option C* is correct.

34. *Option D* is correct.

35. *Option A* is correct. Always remember that system-wide Splunk configurations are stored in the $SPLUNK_HOME/etc/system/default directory.

36. *Option B* is correct. You need a minimum of three to win a majority in the captain election process. Refer to the Splunk docs for the captain election process: https://tinyurl.com/mvz6dvus.

37. *Option B* is correct. Both the SF and RF are set on indexers.

38. *Option C* is correct. Here's the formula: Number of indexers in a cluster >= RF.

39. *Options B* and *D* are correct. Refer to the *Configuring distributed search* section of *Chapter 7, Exploring Distributed Search.*

40. *Option B* is correct. The default value is classic. The remaining options are supported replication policy types.

41. *Option A* is correct. The default bundle size limit is 2 GB; this can be increased in distsearch.conf on the SH. Refer to the *Understanding knowledge bundles* section of *Chapter 7, Exploring Distributed Search.*

42. *Options A, B,* and *C* are correct. There is no replication phase in Splunk.

43. *Option B* is correct. Typically, data input configs go into the `inputs.conf` file in Splunk.

44. *Option A* is correct. For more options, issue the `splunk help monitor` Splunk CLI command.

45. *Option B* is correct.

46. *Option B* is correct. Refer to the *Windows inputs* section of *Chapter 9, Configuring Splunk Data Inputs*, for whitelist/blacklist settings' syntax details.

47. All options are correct. They are valid data input approaches to data indexing. Refer to *Chapter 8, Getting Data In*, for a detailed overview of all data input types.

48. *Option B* is correct. ... (three-dot/ellipsis) notation works recursively, while * (wildcard) works only on file path segments. There is no such notation as %% in Splunk file and directory monitoring input. Refer to the *File and directory monitoring* section of *Chapter 9, Configuring Splunk Data Inputs*.

49. *Options A* and *C* are correct. Both `props.conf` and `transforms.conf` files are used to define the parsing phase setting. Refer to *Chapter 10, Data Parsing and Transformation*.

50. *Option B* is correct. In the `inputs.conf` file, configure `_TCP_ROUTING = <comma separated list of tcpout groups configured in outputs.conf>`. Refer to the `inputs.conf.spec` and `outputs.conf.spec` files for more details.

51. *Options A* and *C* are correct. The other options don't exist in Splunk. Refer to the *Creating search-time field extractions* section of *Chapter 11, Field Extractions and Lookups*.

52. *Option B* is correct. Index-time fields are stored in the index in which data is indexed.

53. *Option B* is correct. The remaining options are totally incorrect. Refer to the *Creating index-time field extractions* section of *Chapter 11, Field Extractions and Lookups*.

54. *Options A* and *C* are correct. `outputlookup` is a write command. The command specified in *option D* doesn't exist in Splunk.

55. *Option B* is correct. The parsing phase will be skipped.

56. *Option C* is correct. For more details on queues, refer to the *TCP and UDP input* section of *Chapter 9, Configuring Splunk Data Inputs*.

Index

packtpub.com

Subscribe to our online digital library for full access to over 7,000 books and videos, as well as industry leading tools to help you plan your personal development and advance your career. For more information, please visit our website.

Why subscribe?

- Spend less time learning and more time coding with practical eBooks and Videos from over 4,000 industry professionals

- Improve your learning with Skill Plans built especially for you

- Get a free eBook or video every month

- Fully searchable for easy access to vital information

- Copy and paste, print, and bookmark content

Did you know that Packt offers eBook versions of every book published, with PDF and ePub files available? You can upgrade to the eBook version at packtpub.com and as a print book customer, you are entitled to a discount on the eBook copy. Get in touch with us at customercare@packtpub.com for more details.

At www.packtpub.com, you can also read a collection of free technical articles, sign up for a range of free newsletters, and receive exclusive discounts and offers on Packt books and eBooks.

Other Books You May Enjoy

If you enjoyed this book, you may be interested in these other books by Packt:

Data Analytics Using Splunk 9.x

Dr. Nadine Shillingford

ISBN: 978-1-80324-941-4

- Install and configure the Splunk 9 environment
- Create advanced dashboards using the flexible layout options in Dashboard Studio
- Understand the Splunk licensing models
- Create tables and make use of the various types of charts available in Splunk 9.x
- Explore the new configuration management features
- Implement the performance improvements introduced in Splunk 9.x
- Integrate Splunk with Kubernetes for optimizing CI/CD management

Tableau Desktop Specialist Certification

Adam Mico

ISBN: 978-1-80181-013-5

- Understand how to add data to the application
- Explore data for insights in Tableau
- Discover what charts to use when visualizing for audiences
- Understand functions, calculations and the basics of parameters
- Work with dimensions, measures and their variations
- Contextualize a visualization with marks
- Share insights and focus on editing a Tableau visualization

Packt is searching for authors like you

If you're interested in becoming an author for Packt, please visit `authors.packtpub.com` and apply today. We have worked with thousands of developers and tech professionals, just like you, to help them share their insight with the global tech community. You can make a general application, apply for a specific hot topic that we are recruiting an author for, or submit your own idea.

Share Your Thoughts

Now you've finished *Splunk 9.x Enterprise Certified Admin Guide*, we'd love to hear your thoughts! Scan the QR code below to go straight to the Amazon review page for this book and share your feedback or leave a review on the site that you purchased it from.

`https://packt.link/r/1-803-23023-1`

Your review is important to us and the tech community and will help us make sure we're delivering excellent quality content.

Download a free PDF copy of this book

Thanks for purchasing this book!

Do you like to read on the go but are unable to carry your print books everywhere?

Is your eBook purchase not compatible with the device of your choice?

Don't worry, now with every Packt book you get a DRM-free PDF version of that book at no cost.

Read anywhere, any place, on any device. Search, copy, and paste code from your favorite technical books directly into your application.

The perks don't stop there, you can get exclusive access to discounts, newsletters, and great free content in your inbox daily

Follow these simple steps to get the benefits:

1. Scan the QR code or visit the link below

https://packt.link/free-ebook/9781803230238

2. Submit your proof of purchase
3. That's it! We'll send your free PDF and other benefits to your email directly